REDUCE, REUSE, RECYCLE

Bromley Libraries

30128 80044 402 1

Reduce
Reuse
Recycle

An easy household guide

NICKY SCOTT

Illustrated by Axel Scheffler

green books

This colour edition first published in 2007
by Green Books Ltd
Foxhole, Dartington, Totnes, Devon TQ9 6EB
edit@greenbooks.co.uk www.greenbooks.co.uk

Reprinted 2007, 2008 (with amendments)

Text © Nicky Scott 2004–8

Cartoons © Axel Scheffler 2004–8

All rights reserved

Text printed by Cambrian Printers, Aberystwyth, UK
on Cyclus Offset paper (100% post-consumer waste)

ISBN 978 1 903998 93 9

DISCLAIMER: The advice in this book about methods of
reusing and recycling is believed to be correct at the time of
printing, but readers should seek expert or professional advice
if in doubt about any of the recommendations made.

*If you would like to discuss a possible bulk purchase of
this book, please phone Green Books on 01803 863260.*

Contents

Acknowledgements

Amanda Cuthbert, for all those edits! Amy Griffiths, for setting me straight on the electrical issues; Paul Marten, MuRF manager from Exeter City Council, for reading various drafts and giving much of his considerable knowledge; Sam Seward for putting me straight on so many things—and for living the reduce, reuse, recycle lifestyle. Last but not least, John Elford at Green Books for asking me to write the book in the first place.

Introduction

This book will try to help you think about the contents of your dustbin in a new light. Hopefully, every time you go shopping it will give you some pause for thought. You will find yourself asking questions like: 'Where does this product come from? Is it made or grown locally? How many miles has it travelled? Can I buy it loose? Who made it? I hope they were well paid! Will it last? Is it made from recycled or recyclable materials? Can I recycle it or the packaging afterwards? Is it toxic? Is there a better alternative?'

For generations, we have been putting our rubbish out for the council to take away and bury in the ground in huge holes called landfill sites. The problem is that these huge holes are filling up faster and faster with increasing amounts of rubbish.

As a society we are consuming more and more of the world's resources, and generating more and more waste(d) materials. In the UK around a third of the contents of an average dustbin are materials that ideally should be composted. At least another third is paper and cardboard which could be composted (although good clean paper and card can be recycled).

REDUCE, REUSE, RECYCLE

Local authorities are under increasing pressure to cut down on the amounts being disposed of in landfill sites and incinerators. Not only do we now have a landfill tax, but a European directive makes it mandatory for all member States to successively reduce the amounts being finally disposed of in landfill. This means achieving much higher recycling and composting rates. It also means changing our perceptions of what 'waste' is. Most of the 'stuff' we chuck away could be re-used, repaired, recycled or composted. It really is wasted! Future generations will no doubt be 'mining' landfill sites for valuable resources.

As well as trying to think about what to do with our rubbish, we need to think about where all this rubbish comes from in the first place. We can make a huge difference by refusing to buy things that will become a problem when they need to be disposed of. This is the main point of this book. By using the A-Z Guide, you can find lots of alternatives to throwing things in your dustbin.

More consumer items are being made now than ever before in the history of the planet. Many of them have an incredibly

short life, and are quickly discarded into the global dustbin, so please try to buy items that are repairable.

There are more people on the planet than ever before, and we have evolved a culture in which we are urged to consume more and more, with little or no regard for where our goods come from and how they are made. Shopping is now a major leisure activity and industry. And the attraction of a particular product depends to a large extent on how it is packaged.

The growing waste mountain

Back in the 1950s, the make-up of the average dustbin was very different from what it is today. Most people had coal fires, so paper was used to start fires, and clinker and ash was a large proportion of the dustbin; the Clean Air Act changed that. Fifty years ago there was virtually no plastic in the dustbin. There were no plastic 'blister' packs and very little processed food; most items were bought loose and wrapped in paper, and people went shopping with shopping baskets. There were returnable bottles, most people had their milk delivered, and other bottles had a deposit on them—a very useful addition to pocket money, and it kept the streets clear of bottles. Rag and bone men still patrolled the streets collecting rags for papermaking, bones which could be used in making bone china and bonemeal fertiliser, and any old bric-a-brac and unwanted furniture.

But change was on the way. With an ever more throwaway society, the amount we wasted grew, along with changes in the way we shopped and a massive increase in ever more alluring packaging. Reliance on processed food grew, which in turn meant even more packaging to throw away. The dustcarts were just not big enough to keep up with our waste: new carts had crushing and compacting mechanisms built into them.

Bottles were no longer returnable, and plastic was used for more and more items. Newspapers got bigger and bigger, processed food became mainstream, and the waste grew exponentially. The local town tips were soon filled up and county councils took on the provision of landfill. Bigger landfills were dug, and still the waste grew. It became increasingly obvious to some environmentalists that the system would have to be changed—but with landfill being so cheap, it has been difficult to make reforms.

Recycling saves energy, saves natural resources and reduces waste disposal. In a world of confusing messages, where environmental problems seem so huge that we as individuals cannot affect them, one of the easiest ways to have a positive impact on the world is to reduce our individual wastage, and to recycle and compost. In fact, some councils are now starting to introduce charging for people who continue to throw recyclable and compostable material in their dustbins. The carrot approach is obviously the best option but the stick is starting to be wielded in some places for people who consistently contaminate their recycling.

Rebecca Hoskins, the BBC camerawoman who made the BBC programme 'Message in the waves', which showed the devastating effect that plastic has, particularly on the marine environment, tells me this book should include a vital fourth 'R': Re-education. It's only a very short time since which we seem to have become reliant on plastic and become such a throwaway society, and we now need to be 're-educated' so that we all think more about the effect our actions and lifestyles have on our environment – see www.plasticbagfree.com.

This book will help you. Every positive contribution, however small is going to help. Give it a try—you can make a difference!

Chapter 1

Away with waste!

Reducing waste is the ideal option. It is staggering how much of what we throw away is perfectly good. A classic example of this is the increasing number of computers that are thrown out by large offices. When a whole system gets upgraded, often all the old ones are thrown into a skip. Yet old office computers can be wiped with approved software and sold on to people on low incomes, or community groups etc.

Fashion also contributes to this waste: in the race to acquire the latest hi-fi equipment or electronic gizmo, and dress in this year's fashionable colours. Some designers are now promoting 'anti-fashion', as a design concept that avoids temporary fashionable styles and promotes quality, durability and less waste.

I recently saw a waste disposal officer talking on a television programme about waste. He said, "It's unbelievable the things that are brought to the Community Recycling Centre. Don't these people have friends and neighbours they could talk to and offer their stuff to? It's a sad reflection on our society that we work to earn money to buy things with, which, particularly at the end of our lives, get buried in a hole in the ground, like us."

Reduce

Buy less stuff! Buy second-hand

Most of what we throw away could be used by other people. Charity shops, boot sales and jumble sales provide a great service in giving our unwanted items another lease of life - but so much more can be done.

> *On average, UK supermarket shoppers spend £470 a year (a sixth of their food budget) on packaging*

Refuse packaging where possible

Virtually every time we go shopping we are offered over-packaged items. Try to source loose items rather than prepacks or blisterpacks packaging. Your local greengrocer or farm shop are much more likely to supply loose and fresher fruit and vegetables, while local butchers typically use greaseproof paper which can be recycled or composted, rather that the plastic containers from the supermarket. In Germany, people routinely remove excess packaging at the checkout for the shop to sort out.

Bag for Life

Instead of picking up those plastic bags at the supermarket and wondering what you are going to do with them, try using a bag for life, preferably not made of plastic. They come in plenty of different shapes, materials and colours, so there's definitely one

out there for you! And they're more gentle on your hands when you're carrying heavy shopping.

Buy quality goods when you can

One way to reduce the rubbish you put in your bin is to buy better quality products. Something really well made by a local craftsperson is going to become something to treasure for life, will help support your local economy and , besides will outlast cheaper, inferior products. Avoid goods that won't last. You can also pick up bargains at second hand shops, furniture reuse projects and auctions to. Why not swap your unwanted items for something you do want? Try www.freecycle.org—it doesn't cost you anything and you can then get something you need, even if it is a bottle of wine or some chocolates. Some designers now talk about anti-obsolescence: meaning designs that are easily repaired, maintained and upgraded so they are not made obsolete by changes of technology or taste. (see 'Cradle to Cradle' in the **Resources** section).

Buy local, think global, and support your local economy

By shopping locally for fresh organic food you might well find that you spend the same amount of money as you would getting in the car and driving to an out-of-town supermarket, and you also avoid packaging which is not recyclable. Generally you will also find you will get better quality food as well! For example compare the rashers of bacon you get from your supermarket to what you find in your local butchers, not only will they be bigger they'll be thicker as well, so plenty more meat for your money! Likewise, the out-of-town DIY store may sell cheaper paint, but are you thinking about how you could have painted half the wall in the time it takes you to get there and back?

• Consider buying local and hand-made products. Support your local artisans. Well-crafted products are the heirlooms and antiques of the future. See www.onevillage.org.

- If you are buying flat pack, mass-produced furniture, look for the type that is sourced from sustainable forest management.
- A lot of flat pack furniture is made from composite materials. This is good, in that some wood recycling is taking place, and less space is taken up during transportation. Old pallets, for instance, can be chipped up and made into new sheet products.

Eat your dinner!

Instead of throwing those left-overs away, why not freeze them to eat another time or have them for lunch the following day? On average a third of the food we buy is just thrown away, which is a waste of resources in so many ways, such as transportation (cars and trucks), preparation (cooking) and when it ends up in landfill. It could also save you money on your food bill to! Have a look at www.lovefoodhatewaste.com which has some great tips about keeping food fresh and some great recipe tips too.

Avoid disposables

Try to avoid items that are used once and then thrown away—especially nappies (see the nappies section in this book for more information about the alternatives), but also razors, cameras, plastic cutlery, cups and plates and so on. For example for parties and events, you can hire cutlery, cups and plates instead of buying disposables.

Avoid anything you can't reuse or recycle, where possible

For instance, many household and garden chemicals should not be disposed of in the dustbin. If you use materials which can be recycled or composted, then you aren't left wondering what to do with a toxic substance. (See the garden chemicals section for more information.)

Refill

Wherever possible, buy products in refillable containers. These include a range of soap products, washing-up liquid, washing machine liquid, multi-surface cleaner, cream cleaner etc. These items are most commonly found in local health food shops, check with your local store if they offer this service.

Some food shops also do a wide range of loose food products (and they're often cheaper), so that you can take what you need with minimal packaging - even bring your own. There is a great new store in London called "Unpackaged" which sell everything loose and even provides reusable containers (at a cost,) if you forget to bring your own. Check out www.beunpackaged.com for more information.

Buy in bulk

If you buy some of your food in bulk, you can:

- Decant or bottle up smaller quantities for use (which is cheaper in the long run)
- Band together with friends and bulk buy your food—this will reduce the cost for all of you
- Have fun being creative with your bulk buys, such as marinating your own olives in oil with herbs and spices

Reuse

We are used to throwing things away without a second thought, but we can:

Repair Furniture is often thrown away when it could be given a whole new lease of life in the hands of a local craftsman.

Reinstate Chairs can be re-caned, sofas and chairs re-upholstered.

Reactivate A rusted-up chain on a bike can often be rejuvenated by soaking in penetrating oil, as can other rusted-up machine parts. Of course, bicycle chains would not have to be reactivated if they had been regularly oiled in the first place!

REDUCE, REUSE, RECYCLE

Reinvent Find new uses for everyday items: e.g. plastic drinks bottles make great mini cloches to protect tender plants from slugs. Or a clean plastic takeaway container makes a great food container.

> *In the UK we have gone from zero recycling to a national average of 13% in a very few years.*

Donate Could charity shops, jumble sales, hospitals, playgroups, residential homes etc make use of your unwanted things?

Many charity shops are now becoming extremely good at marketing a whole range of products. The CDs which used to be put in a box on the floor, are now being displayed as they would be in a music shop. Some charity shops are becoming specialists in particular areas: designer clothes are on sale in several London charity shops, whereas others specialise in 'retro', with old telephones selling for £30 alongside 1950s and 1960s china and furnishing fabrics. Some specialise in books, others in furniture and so on. Browsing in charity shops is becoming positively trendy!

In Amsterdam, people put useful items out on the pavement on the evening before the refuse collection, and whole streets turn into free flea markets. In London anything left on the pavement soon disappears. It even works in the small town where I live, so I'm sure it's an idea that could catch on anywhere. And now you can freecycle—join the Freecycle community nearest you: www.freecycle.org and www.swapitshop.com are websites where you advertise your unwanted items for free, and in return you can ask for something you want. Also see www.readitswapit.co.uk for books.

So much of the stuff that gets taken to recycling centres or thrown out can be sold. Car boot sales are brilliant for this—make some money too! Auction houses will also sell anything from prime pieces of furniture to miscellaneous boxes with the most unlikely assortment of odds and ends. And nowadays there are auction sites on the internet, such as eBay.

Sell So much of the stuff that gets taken to recycling centres or thrown out can be sold. Car boot sales are brilliant for this—make some money! Auction houses will also sell anything from prime pieces of furniture to miscellaneous boxes with the most unlikely assortment of odds and ends. And nowadays there are auction sites on the internet, such as eBay.

Recycle

Recycling in this country is still in its infancy. We have a long way to go, but soon everyone in the country will be offered a doorstep collection for at least some of the materials that could be recycled. Some parts of the country that have been recycling for longer have achieved far higher recycling rates than the average. They are now moving into extending the range of materials recycled, for example by collecting kitchen and garden waste for composting, and plastics. Don't use that plastic box from the council to store toys or tools: it's for recycling! Most of us now have some sort of doorstep scheme. Read the information that comes with it, or contact your local council for more information. If you have internet access, look on your local council's website.

Get into the habit
More and more councils are offering kerbside collections for recyclable materials, encouraging us to make recycling a habit. You can really help your local council by checking what they will or will not accept for recycling in your area. Different councils operate different systems: some have bags, others use boxes or wheelie bins. If you have a recycling box it helps to put cans, card, and newspapers/magazines into separate bags within the box.

Clean rubbish
Efficient recycling depends on having clean, separated material, which doesn't need sorting before it is processed. For instance

putting china in with glass bottles can lead to whole loads of glass being rejected. This happens when Pyrex, plate glass and drinking glasses are mixed in with glass bottles and jars. It is becoming a part of our daily life to rinse out our bottles, jars and tins for recycling. Most of us have easy access to at least the more common recycling banks for glass, newspapers and magazines, tins, cans and cloth es. So please remember to put your material in the right place, and take note of what the signage on the container states can and cannot be accepted.

Recycling bulky items—the skips!

Councils are busy improving their sites where householders can take 'bulky' household waste. Anything from old sofas, bikes, baths and bricks to woody prunings, branches, plastic bottles, glass, waste oil and timber. These facilities are undergoing a face-lift so that more and more is saved rather than being buried or burnt.

Contact your Local Authority to find out where your nearest recycling facility is, and what you can or cannot take there. Alternatively, you may have a local community-operated scheme—if so, contact them first!

Buy Recycled

Purchase items made from recycled products: they are becoming better all the time, as more people are demanding them.. Many companies in all areas of our lives are offering items made from recycled materials. However be careful, if a label says "environmentally friendly" it is not necessarily made from recycled materials. Purchasing a product with "100% post-consumer waste" on the label means you are guaranteeing that its content is made of recycled materials. See www.amazingrecycled.com. Find out whether your workplace has a procurement policy to buy recycled products, and if not, suggest it!

Compost

Home composting is great—it's just about the only reprocessing that we can all do ourselves and it removes such a lot of waste from the dustbin. Nearly two thirds of our rubbish consists of material that could be composted, and much of that is paper and card. It is also a great place for uncooked food such as peelings, skins, egg shells and even tea bags! For tips and advice have a look at www.recyclenow.com/compost. It's staggering how much food is wasted each week in this Country – a charity called 'Fareshare' www.fareshare.org.uk is addressing this situation by collecting 'short shelf life' food from supermarkets and other food outlets to be distributed to needy people. We can make sure that we do not thoughtlessly throw perfectly good food away and that we develop composting systems for the rest. Schools are rediscovering the joys of composting so hopefully a new generation will take it as second nature. Some schools are now discovering that if they let children help themselves to their own food that there is far less food wasted. Many schools are now composting all their food waste with specialist composting equipment. In fact anywhere that has the space to compost and to use the resulting compost can now compost food waste in situ and save themselves money in the long run. Advice is available via the compost doctors: www.compostdoctors.org.uk.

> *In the UK we produce enough rubbish every day to fill Trafalgar Square to the top of Nelson's column, and enough to fill the Albert Hall every hour.*

Chapter 2

Some recycling myths

There are a number of arguments which are all too often directed at recycling, for example that it uses more energy and is more expensive than using raw materials. This is untrue (see below).

Critics argue that recycled materials such as glass are transported hundreds of miles to be melted down. In rare cases this does happen, often because of a lack of local facilities. However glass made of raw materials is also transported from where it is produced, and this uses more energy (recycling just one glass bottle saves enough energy to power a computer for twenty minutes,) and mining the raw materials has a significant impact on our environment.

The Myths

1. "There's no point in recycling, because all the stuff just gets dumped."
This is a common story in the press. What is often not realised is that if too much of the wrong material is found in a container it will contaminate the whole load and make it unusable. This is because of the cost of sorting the load. The vast majority of loads are fine and end up being reprocessed, but unfortunately bad media stories tend to stay in the mind longer than good ones.

Recycling is worthwhile, and is getting better all the time, as the technology surrounding it develops and evolves. Also more and more people now put out clean, sorted material for recycling. A visit to a Materials Recycling Facility or 'MuRF' is fascinating. MuRFs can sort plastics out mechanically or by hand. In some you can see completely unsorted materials coming in. The contents travel along conveyor belts where they are tumbled through huge revolving drums, passed under powerful magnets, blown with air knives and zapped with laser beams. These beams identify the types of plastic which are blasted with jets of air down the appropriate chute. These processes perform the major sorting out of materials, light from the heavy materials , the steel from the aluminium, and the small from the large.

2. "It costs more to recycle than to make things from new materials."
This is not true. Many recycled products not only save energy and water but also reduce raw material usage and the associated energy and pollution caused in the process of obtaining the raw materials. For example, it is far better to be constantly recycling aluminium than wastefully mining out the finite stocks of bauxite, and causing unnecessary pollution and wasting energy in the process.

3. "It costs more to recycle than it does to throw rubbish away."
Rubbish collection costs us all money, but the real cost has been disguised due to subsidised landfill costs (we have the cheapest rubbish disposal costs in Europe). With landfills filling up and closing down, and landfill taxes having been introduced and increasing every year, recycling makes more and more financial sense. Councils will soon be fined £150 per tonne of material over and above their mandatory target.

Manufacturing 1 tonne of recycled paper results in 74% less air pollution and 43% less water pollution compared to the manufacture of paper using virgin wood pulp.

4. "It causes huge amounts of pollution trucking all the recyclable materials around the country."
It is true that there are relatively few reprocessors for recyclable materials at the present, and that mileage costs and associated pollution have to be taken into account. However, as more uses and markets are developed and more local reprocessing takes place, the market price for recycled materials will rise.

5. "Washing out tins and bottles uses more energy than is saved by recycling them."
Not if you wash them out at the end of doing the washing up!

6. "It all takes up to much time, and I don't have enough space."
Recycling involves dedicating a container for recyclable materials. In the long run you are likely to save space in your rubbish bin and should therefore need a smaller bin; leaving you room for a container for your recyclables. Or you could have a small

container in the kitchen, and feed this into your kerbside bin/box or container for the recycling centre. It's all about routine, once you are into the swing of recycling, it takes very little time, and you won't even notice you're doing it!

7. *"You have to take all the labels off—it's all too much trouble."*
Labels don't have to be removed from bottles or tins, but food containers do need to be rinsed out.

8. *"There's so much green glass that it just gets thrown away."*
Only contaminated loads of glass ever get thrown away (see 1 above). As new uses are being found for materials like glass it creates a new and more competitive market place. There are now dozens of alternative uses for glass—see 'glass' in the **A-Z Guide**.

9. *"Recycled products are poor quality."*
Products made from recycled materials are every bit as good as (if not better than) products made from raw materials. For example, throughout the world, military and commercial aircraft use retread tyres, and this is in an area where safety is paramount. See www.greenshop.co.uk.

Aluminium, glass and some plastics can be recycled indefinitely without a loss in quality. Paper, however, can only be recycled a limited number of times because the fibres get shorter, but it can be reprocessed into other products. Many paper processors mix fresh, long fibres with reprocessed paper to make new paper.

10. *"Recycled products are too expensive."*
Not always! However, the market is new, and creating demand for products using recycled materials is important and will help to lower prices in the longer term.

Chapter 3

A—Z Guide

Aerosols

REDUCE / AVOID CFCs (chlorofluorocarbons) are now not used in aerosols, but can persist in the atmosphere for 50 to 100 years. Although the holes in the ozone layer now seem to be mending, it will take another 40 to 50 years to see if it has recovered. 75% of local authorities now collect aerosol cans.

RECYCLE Can be recycled when empty, but check with your local authority first (you can put your postcode into the recyclenow website, www.recyclenow.com). DO NOT SQUASH or pierce, as they can explode.

Aluminium

RECYCLE Wash and squash cans first. Aluminium is the most valuable of our commonly recycled materials, and is one of the most important items to keep out of your dustbin.

SELL *or* **DONATE***:* Large items such as cooking pans, cutlery, window frames etc can be taken to a scrap dealer, your local charity shop or recycling centre.

Aluminium Foil and Containers

REDUCE Aluminium foil is useful stuff, but you don't have to use it to wrap everything in. Why not use a reusable container instead?

REUSE it as much as possible, clean, flatten and put it back in the drawer for next time. Containers can be used for seed trays.

RECYCLE THE SCRUNCH TEST is a test to check if your foil is the real thing. Scrunch it into a ball: if it's foil it will stay crushed in your hand, but plastic will spring back! Many local and kerbside schemes will collect it for recycling. It needs to be collected separately from aluminium cans and pans (as it is a different alloy and has different end uses), but it can still be reprocessed. Some charities and special appeals will take it, for example www.alupro.org.uk.

Antifreeze

RECYCLE DONATE: Give it to your neighbour or local mechanic.
DISPOSE It is hazardous waste, so take to a recycling centre.

Appliances—*see Electrical & Electronic Appliances*

Asbestos

Asbestos is a very dangerous substance when disturbed in any way: contact your local council for advice or look at the Direct Gov website: www.direct.gov.uk. Do not attempt to remove or repair asbestos without getting advice from an expert. Do not take asbestos to a recycling centre without contacting them first to see if they can accept it.

Asbestos Fire Blankets—*see above*

Ash

REUSE the clinker for garden paths, or make a narrow path across your lawn for the winter.
RECYCLE *Compost:* Wood ash can be sprinkled lightly into your compost heap or around your garden. It is also used by potters in glazes and in soap making.
DISPOSE Coal ash should be disposed of in your dustbin once it has cooled.

Attic Clearance

Think first! What can you sell or donate or recycle? Use local charity shops etc. www.ebay.co.uk www.freecycle.org and www.swapitshop.com.

Autumn Leaves

RECYCLE / *Compost:* Put wet autumn leaves in a plastic bag, stab a fork in to make some holes, and a year later you'll have leafmould. Or your can take them to your local Recycling Centre. Even better, pick the leaves off the lawn with a mower which will mix grass cuttings and chop the leaves up, and you'll have leafmould even more quickly. Don't burn leaves, as they produce highly carcinogenic smoke—particularly bad for babies and children—and it's a waste of all that lovely material. Check out www.homecomposting.org.uk/content/view/19/34/ for more information, or my books *Composting for All* and *Composting: an easy household guide*.

Baby Goods

RECYCLE SELL, DONATE *or* **FREECYCLE***:* Good condition clothes and equipment can go to second-hand shops and charity shops. Contact your local playgroup or nursery school—put up a card on their noticeboard.

Bags—*see Plastic Bags*

Bathroom Fittings and Furniture

REUSE Sinks, soil pipes and even old toilets (and chimney pots) can be used for plants: a sink in the ground is useful for containing mint and other invasive plants.

SELL *or* **DONATE***:* Advertise locally, sell to architectural salvage yards or give to recycling centres.

RECYCLE Broken ceramic fittings can go with rubble and brick for recycling.

Batteries (Car)

RECYCLE Car batteries are routinely recycled in the UK, with a current recycling rate of approximately 90%. They are collected at garages, scrap metal facilities and many civic amenity and recycling centres.

Batteries (Household)

> *The UK generates 20,000—30,000 tonnes of wasted general-purpose batteries every year, but less than 1,000 tonnes are recycled.*

With so many appliances being battery-powered, it really makes sense to seek out alternatives, such as solar-powered, or rechargeable batteries. At present only about four percent of batteries are recycled (about 1,000 tonnes) – that means about 24,000 tonnes of batteries aren't! The energy needed to manufacture a battery is on average 50 times greater than the energy it gives out. www.cat.org.uk www.freeenergynews.com.

RECYCLE Some manufacturers will accept spent nickel-cadmium batteries if returned to them. Ask your local Council if they are providing a service yet, since the first edition of this book came out many more are.

See www.envirogreen.co.uk > *services* > *battery*
www.wasteonline.org.uk & search for 'batteries'
www.letsrecycle.com/equipment/batteries.jsp

Why recycle batteries? Whilst the exact chemical make-up varies in different types, most batteries contain heavy metals which are a cause for environmental concern. When disposed of incorrectly, these heavy metals may leak into the ground when the battery casing corrodes. This can contribute to soil and water pollution, and endanger wildlife. Cadmium, for example, can be toxic to aquatic invertebrates and can accumulate in fish, which makes them unfit for human consumption. Some batteries, such as button cell batteries, also contain mercury, which has similarly hazardous properties. Mercury is no longer being used in the manufacture of non-rechargeable batteries, except button cells where it is a functional component.

Bedding and Blankets

RECYCLE SELL *or* DONATE: Charity shops will sell clean, good quality bedding etc. Reuse for rags, dust sheets etc, and many animal rescue centres will also accept them to reuse—see **Textiles**.

RECYCLE COMPOST: Pure wool, cotton, linen and other natural fibres can be composted. However, you cannot compost man-made fibres, or natural materials if mixed with man-made fibres.

Beds

RECYCLE SELL *or* DONATE to furniture projects, social services or your local recycling centre. Mattresses can only be reused if they conform to current fire safety standards. One mattress takes up over 23 cubic feet in a landfill site; a project in Scotland called 'SpringBack' disassembles mattresses (and beds) so that the component parts can be recycled. Find your local furniture reuse project through www.frn.org.uk.

Barrels *(Beer, Wine etc)*

REUSE Contact the original supplier of the barrels: they can be re-used. If too far gone to hold liquid (the barrel, not you) then *Sell* or saw in half to make plant containers. You can also make a 'solar composter with, preferably a 'two-thirds' barrel – see www.smartsoil.co.uk & click on the sunfrost page.

Beverage Cartons—*see Cartons*

Bicycles

RECYCLE metals for scrap value.

SELL *or* DONATE: Advertise locally or use auction sales, jumble or boot sales. Bicycle shops will often refurbish your bike for resale or dismantle for spares.

Bicycles are collected by Re-Cycle, the bicycle charity. The bicycles are sent to developing countries where they are repaired and reused. See www.re-cycle.org and www.bikerecycling.org.uk for more information.

Birthdays—*see Christmas*

Blockboard

REUSE Take to a recycling centre for re-use. You should not burn composite woods like blockboard because of the resins and glues, which are toxic.

RECYCLE Blockboard can be chipped up to make chipboard etc.

Books

REDUCE Use your local library.

RECYCLE Some councils and charities also have book banks for recycling books back into paper.

SELL *or* **DONATE**: Books in good condition can often be sold to second-hand bookshops—otherwise donate to charity shops etc.

www.readitswapit.co.uk does what it says – use the site to swap books.

Bottles & Jars

Almost 200 glass jars and bottles are thrown away in Britain every second.

REUSE Try to buy reusable/refillable ones. Jars and bottles also have many reuses, apart from recycling. Lids can be attached to the underside of shelves, and the jar can be screwed in having been filled with nails or screws etc. And they can be used for all manner of small amounts of foodstuffs, paint, etc.

REDUCE, REUSE, RECYCLE

RECYCLE Remember to remove the lids: if you mix metal and glass, the load could be rejected and sent for landfill. The lids can go with the tins and cans.

Bric-a-Brac

RECYCLE SELL *or* **DONATE***:* Save for jumble sales, charity shops, car boot sales, freecycle etc.

Bricks

REUSE Can be cleaned for re-use. Bricks have masses of uses in the garden: supporting water butts, troughs etc, for paths, raised beds and of course for new building projects.

RECYCLE at a Recycling Centre.
SELL *or* **DONATE***:* Advertise locally.

Brochures—*see Catalogues*

Brushes

REDUCE Paint brushes last longer if you look after them:
- Buy good quality paintbrushes and clean them carefully
- If you are going to use them again soon, put brushes from oil-based paints (check the tin) in water to the top of the bristles to stop them drying out. Load emulsion paintbrushes and rollers up with plenty of paint and put in a pot (or tray) with a plastic bag over the top..
- When you have finished, wipe brushes on newspaper first— use a small amount of white spirit for oil brushes and then use more newspaper. For emulsion brushes, use hot water and a spot of detergent.
- Keep brushes used for black paint just for black paint jobs.

REUSE Toothbrushes can be reused for all kinds of cleaning jobs. Trim the ends of any gummed up brushes with scissors for a new lease of life.

Bubble Wrap

REUSE It's very therapeutic to pop it! However, local potters and craft shops etc can reuse if it's big enough, and unpopped! Wrap plants and plant pots in it to protect from frost. Use it for transporting breakable items, when moving house, or for wrapping things before sending them through the post.

RECYCLE Check whether your local authority or civic amenity site is collecting polythene, which includes bubble wrap, as some now are.

Building Materials

REUSE SELL *or* **DONATE**: These make up a huge part of the waste problem. Any decent quantity is worth advertising in the local paper—or you may have a local community project or reclamation yard. When undertaking building work it pays to set up a clear system so that materials can be separated out: wood, bricks, blocks, rubble, mortar, metals, wiring and cables, plastics, windows and doors etc. Also see under individual headings.

If you are renovating your house:

- Bricks can be reused, e.g. in the garden
- Any large pieces of stone can be incorporated into garden design
- Metals and cables can be sold to a scrap dealer
- Doors and windows can go to an architectural salvage yard
- Wood can be reused or burned
- Mortar and old plaster can be bagged up for reuse in the garden.
- Rubble can form the base of paths or tracks.

Modern houses tend to be made with many more man-made materials which are not easily recycled: PVC windows, doors, gutters etc, composite woods bound with resins, foam insulation and so on. When these buildings are demolished in the future, there will be fewer useful recyclable or reusable materials.

For reclaimed building materials see www.salvo.co.uk.

Building Rubble

RECYCLE **SELL** *or* **DONATE**: If you are planning any building projects you will need some rubble for hardcore: store it in a corner until needed. It is also useful for landscaping foundations in the garden too. You could advertise locally: give it away free to anyone willing to take it away, or as a last resort take to a recycling centre.

Bulky Refuse

RECYCLE Most councils offer a bulky household collection service, but make sure it really is unwanted or unusable by anyone else first before you contact them. Join www.freecycle.org!

Cameras—*see Disposable Single-Use Cameras*

Cans—*see also Aluminium*

RECYCLE Rinse cans at the end of your washing up, and recycle. Aluminium cans are the most cost-effective materials to recycle, so don't waste them! Recycled aluminium from cans is used extensively to make new products.

Steel cans are also valuable and well worth recycling. They are easily separated from the aluminium with magnets at the Materials Recovery Facility (MRF). On average, every single person in Britain uses 240 steel cans each year.

Every day 80 million food and drink cans end up in landfill.

Cardboard

REUSE Large cardboard sheets are useful in the garden as a weed-suppressing mulch. Ideally, cover them with compost.

RECYCLE Shred it for animal bedding. Many recycling schemes will collect or accept cardboard or you can take it to your local Community Recycling Centre.

Compost: Cardboard is great for compost heaps—worms love it! Either line your heaps with it or rip it up, or just put in layers with wet green material like grass cuttings—see also **Compost** section below. If adding a lot, it's probably best to wet it first.

Cardboard tubes: Apart from composting or recycling, cardboard tubes have many reuse applications. Use them as storage for, brushes, cables – all sorts! Small pets also find them great toys! Please make sure any plastic or metal is removed first. Shigeru Ban, a Japanese architect (see www.shigerubanarchitects.com) has designed a clever way to link together cardboard tubes so that a roll-up screen can be made from them.

See www.recyclethis.co.uk and www.junkk.com for loads of ideas.

Cards

REUSE Card can easily be reused by cutting the picture off for a new card, pasting clean paper over the existing message or using the picture or logo as a gift tag

RECYCLE Greetings cards are recycled along with cardboard by some councils, so check to see if your council will accept them.

DONATE: Some shops and some council offices have boxes for Christmas cards (see also **Christmas**).

Carpets

REUSE Make a carpetbag!

Clean Many carpets are disposed of simply because they are dirty! Hiring an industrial-strength steam cleaner for the weekend can transform your carpets, and kill any clothes moths that may have taken up residence. If the moth has taken hold, woollen carpets and felt underlay can be used as a mulch material in the garden or allotment. There are many companies who specialise in carpet

cleaning and refurbishment if you don't want to do it yourself.
Replace Another option, common in offices, is to use carpet squares, which can then be replaced individually as necessary.
SELL *or* **DONATE**: Offer them to social services, hostels, local schools, scrapstores etc, or advertise locally.

Recycled carpets Recycled carpet can be moulded into hard durable 3-D forms, or processed into a hard board material, by combining heat and pressure. The carpet then assumes new properties and becomes waterproof, oilproof and highly durable, with a smooth surface. See www.carpet-burns.com for more details.

Lease your carpets! Businesses now often do not own their carpets—rather they lease floor covering by the square metre. See www.interfaceflor.eu.

Carrier Bags—*see Plastic Bags*

Cars & Car Spares

RECYCLE **SELL** *or* **DONATE**: Failing that, you should always take end-of-life vehicles to a proper breaker's yard, where useful parts are removed for re-use or sale. Old and unusual vehicles will usually have their own dedicated clubs who will often take your vehicle and lovingly restore it. Useful spares can be advertised or your local garage may well be happy to take them.

See www.zyra.org.uk/scrapcar.htm.

Cartons

REDUCE Avoid where possible. Milk, fruit juices etc can all be purchased from your local milk man in glass bottles, which can be refilled up to 13 times before they are recycled into new glass. To find your local milkman, check out www.milkdeliveries.co.uk.

REUSE Cartons can be used as plant pots—on their sides for seedlings, and upright for single plants.

RECYCLE Tetrapak cartons are difficult to recycle as they are made of plastic, foil and cardboard. However, a number of councils now have collection schemes and 'bring banks' for cartons. To find out where they are located contact your local council. To see how they are recycled, take a look at www.tetrapakrecycling.co.uk. To see how juice cartons are turned into mouse mats and clipboards, see www.cutouts.net.

Cartridges (ink jet, laser etc)—*see Toner Cartridges*

Catalogues

RECYCLE They can go with newspapers and magazines for recycling.

Cat Litter

RECYCLE *Compost:* Cat litter can be added to the compost heap—however you must be aware that cat faeces can contain parasites, the composting process is a better route than burying or disposal but you must give it time, a hot heap will work more quickly. A green cone will digest your cat litter (along with all food waste,) and release all the products into the soil in your garden. Have a look at www.greencone.com for more information. Other enclosed systems will also do both see www.smartsoil.co.uk for a rotating insulating compost bin that will really heat up your food and pet waste.

You can now buy cat litter made from recycled paper, hemp and mineral sources—for example bentonite clay.

CDs & DVDs

REUSE
- Hang in the garden or allotment as bird scarers
- Use as reflectors in your drive
- Use as coasters for drinks
- Sell to record shops
- Donate to a local scrapstore or charity shop for reuse

See www.cutouts.net for more reuse ideas

RECYCLE Send to The Laundry CD Recycling, London Recycling, 4d North Crescent, Cody Road, London E16 4TG. For other contacts for recycling see www.reuze.co.uk/cds.shtml.

Cellophane

Genuine cellophane is derived from cellulose and is biodegradable. Some food products are still wrapped in it, but most so-called cellophane is in fact polypropylene and not biodegradable. The only way to tell is either to contact the manufacturers or to try to compost it. See **Envelopes** for where to buy genuine cellophane window envelopes. See www.pak-sel.com/sub1.htm.

Ceramics and China

REUSE Broken china can be utilised in mosaic work. Mosaics are becoming more common in civic spaces. Ceramics can also be ground up for incorporation into new building blocks. Offer it to a local art school/college, who can use it in their work. Put up a card at your local crafts centre, or advertise locally if you have a quantity of interesting broken ceramics.

Never put ceramics, old drinks glasses or Pyrex in with glass for recycling, as it will contaminate the entire load and will have to be landfilled. Put it in your rubbish bin if you don't want to reuse it.

Some incredible work has been produced using ceramics: see the work of Gaudi in Barcelona, where there are whole buildings covered in ceramic pieces, or Niki de St Phalle's Tarot Garden in Tuscany , and the work of Nek Chand in India. www.nikidesaintphalle.com www.nekchand.com.

Cereal Boxes

RECYCLE Your Local Authority may collect cardboard.
COMPOST Flatten then rip or scrunch before adding to the heap.
REUSE Playgroups and nurseries often collect cardboard tubes and boxes for 'creative play'. See also **Children**.

CFCs (Chlorofluorocarbons)—*see Fridges & Freezers*

Chemicals—*see also Garden Chemicals*

By chemicals we generally mean hazardous or toxic substances used in the household which include paint, white spirits, bleach, antifreeze, brake fluid, engine oil, garden and household chemicals, woodworm treatments and so on. Some of these are listed in this A-Z Guide, but if in doubt contact your local authority for disposal advice.

REDUCE Try to find alternatives to the most toxic ones—check the labels.

DISPOSAL Never dispose of chemicals down the sink or drain. Unwanted pesticides and other garden chemicals must be disposed of properly—all hazardous substances should be taken to a recycling centre for safe disposal. Contact your local council first for advice or phone the Environment Agency ☎ 08708 506 506. www.environment-agency.gov.uk.

Since December 2003 it has been illegal to have or use a wide range of chemicals in the household. These include creosote, and several of the older pesticides which may be lurking in the back of the garden shed. See www.nhhwf.org.uk for more information or contact your local council.

Cleaning

Avoid harsh chemical cleaners: you can use cleaners (for all household uses, including yourself and your clothes) made from natural products and even use a steam cleaner—then you won't have to dispose of any hazardous chemicals.

See www.greenshop.co.uk or www.naturalcollection.com.

Children

I'm not suggesting that we should recycle children! Children really like the idea of recycling (and composting) and it's never too early to introduce them to it.

REDUCE, REUSE, RECYCLE

If you are a teacher or group leader and want to do more, then why not contact your local authority and see if they are working with schools, and try Wastewatch www.wastewatch.org.uk or Global Action Plan www.globalactionplan.org.uk.

A lot of the materials we throw away are essential components in 'Blue Peter' type constructions. When children are feeling creative, it's great to have boxes of materials to hand, for example:

- cardboard boxes and tubes, plastic bottles and yoghurt pots plus glue, string and silver foil
- jam jars (for older children) and tins with lids can hold small items like buttons, bottle tops, nails, screws, odds and ends
- Pieces of material with thread, cord, ribbons and wire for fastening things.
- Paper used on one side and pieces of card, for painting and glueing pictures to
- Magazines and catalogues with pictures of animals, trucks, landscapes, flowers etc, to cut up for collages

If you find you're collecting too much, try offering it to your local school, playgroup or nursery.

For a bit of fun, try putting a crisp packet in the oven on a low temperature—it shrinks to a fraction of its full size. It's fascinating to see a miniature of the product with all the writing barely legible. See also **Scrapstores**.

Chipboard

REUSE Keep dry for re-use, as otherwise it will swell up and be useless.

DISPOSE Chipboard contains toxic glues and resins, so do not burn: dispose of small bits in a dustbin, or take to a recycling centre for disposal.

Christmas

REDUCE You can reduce your Christmas waste by:

- Trying to really think about the presents you buy, and sourcing locally and ethically
- Buying cards made from recycled card
- Supporting your local shops and craftspeople
- Reusing good quality wrapping paper
- Recycling old Christmas cards: there are often collection points in Post Offices, council offices or shops.

REUSE You can cut up Christmas cards for re-use as either new cards or gift tags. Most Christmas decorations get reused year after year.

RECYCLE Don't put cards in the newspaper and magazine recycling collection as card fibre cannot be recycled into paper, but it can go in any cardboard collection you may have. Cards can also be recycled at your local Recycling Centre in the Mixed Paper container. Alternatively there are many major stores collecting cards for charities such as the Woodlands Trust. (see cards).

Paper chains and other paper and card decorations can go for recycling along with cardboard, but broken glass balls should be wrapped in paper and put in the dustbin. For more ideas go to www.mindfully.org and search on Christmas—it's a great site!

Christmas Trees

REUSE Why not buy your Christmas tree this year with the roots? You can then plant it in the garden and have a bigger tree each year!

RECYCLE *Compost:* Many councils now offer Christmas tree collection points. The trees are shredded for composting or used for stabilising sand dunes. You may have a local community composting project which will take it—see Community Composting Network (CCN) in the **Resources** section.

Cleaning—*see Chemicals*

Clingfilm

REDUCE Clingfilm is useful stuff, but there's no need to wrap everything in it! Why not use a reusable container to put your sandwiches in? It stops them getting squashed to! You can also use the containers to put food left-overs in the fridge or the freezer too. Putting food in a bowl in the fridge with a plate over it works fine!. Clingfilm cannot yet be recycled.

Clothing—*see also Textiles*

REDUCE The fashion industry wants us to constantly change our clothes, and huge quantities of clothing are discarded as a result: do you really need that new jumper?

REUSE
Sell or Donate: Good clean clothing can be reused: take it to jumble sales, charity shops, or put in clothing banks, which can be found all around the county, such as at your Local Recycling Centre. Also:

- Some businesses make completely new clothing from old e.g. www.traid.org.uk
- Children love dressing up both themselves and their dolls and teddies—keep a box of clothes handy.
- Contact your local amateur dramatic society

RECYCLE Some scraps of material are wanted by quiltmakers. If you have even quite small pieces of (especially) small print material, they will put it to good use. Advertise locally or in a craft shop, or contact the Quilters Guild. www.quiltersguild.org.uk.

What happens to my old clothes?

When you put old clothes into a textile bank:

- 70% is used as second-hand clothing and shoes to meet demand in both the UK and developing countries
- 9% is reclaimed, shredded and used as a filling material (e.g. for insulation or stuffing mattresses)
- 8% goes to the reclamation of fibres to produce recycled products
- 7% becomes wiping cloths for industry, replacing paper and equivalents
- 6% is rejected as waste (bags, zips etc).

See www.traid.org.uk
www.wasteonline.org.uk

Coat Hangers

REUSE DONATE Charity shops and dry cleaners often need hangers.

RECYCLE Metal coat hangers can be recycled in your Community Recycling Centre's metal container, while plastic coat hangers can be placed in the bulky plastics container, if your council has one.

Coffins—*see Funerals*

Coins

REUSE AND **RECYCLE**
DONATE: British and Foreign coins can be given to charities. Hand in at your local charity shop, e.g.:

- Oxfam: www.oxfam.org.uk
- The Royal National Institute for the Blind: www.rnib.org.uk (or ☎ 020 7388 1266 for details of your local RNIB office)
- Banks, post offices and other shops sometimes collect foreign coins for recirculation or to give to charity

SELL to collectors or to specialist shops.

Compost

REDUCE Make it yourself, but if you have to buy it, buy it locally. You may be lucky enough to have a community composting group or composting business in your area who will be able to take your garden waste.

Don't buy peat-based composts or growing medium. See **Peat**.

RECYCLE The growing media from old hanging baskets, grow bags etc, can be spread around the garden or added to a compost heap.

To find out more about composting, see *Composting: An Easy Household Guide* in the **Resources** section.

Computers

REUSE Many recycling projects and commercial companies will take computers, monitors and associated hardware. They can wipe hard drives with special software and re-sell equipment to low income groups etc. Even broken equipment can be taken apart for components. Contact the Community Recycling Network (see **Resources**) for contacts.

If you work in an office, tell your manager about these services before your whole system is upgraded and the old computers junked.

SELL *or* **DONATE**: www.donateapc.org.uk Computer Aid Int'l, 433 Holloway Road, London N7 6LJ ☎ 0207 281 0091. www.computeraid.org www.recycle-it.ltd.uk ☎ 0870 774 3762.

RECYCLE Since the WEEE (Waste Electronics and Electrical Equipment) Directive came in to force in July 2007, all electrical items must now be recycled. You can take your computer to your local Community Recycling Centre, or the company you purchased it from has a legal obligation to take it and recycle it. N.B. Ensure all personal information has been wiped from the hard drive beforehand.

Cooking oil

REUSE Bio-diesel projects are springing up around the country—imagine powering your car on your old chip oil! If you have a local community recycling project, they should be able to tell you more. See an example of a biofuel project at www.rapoleum.com.

Old oil can also be used to add life to wooden garden furniture, wooden compost bins, etc (thin with paraffin first). Also, try using a container with sand and oil which you clean garden forks and spades in: it cleans and oils at the same time.

RECYCLE *Compost:* If you really can't find a use for it, then oil can be composted. It's best to mix well with paper or cardboard first. www.biodiesel.co.uk.

Copper

RECYCLE SELL Copper pipes and hot water cylinders are often dumped, but they are valuable and can be reused or reprocessed. Take to a scrap metal dealer or council recycling site.

Corks

REUSE Corks are not being recycled at the moment, but their time will come! In the meantime they have some re-uses:

- You can use them for lighting fires or for soundproofing (they're sometimes used in pubs)
- You can make cork noticeboards from whole corks by glueing them together in a frame.
- Put them on the end of sharp tools such as stanley knives or bradawls, or on the end of garden canes.

Cork facts

- Cork recycled from wine bottles can be used to make products such as place mats, floor tiles, gaskets, fishing rods, shoe edges and insulation material.
- Real corks are preferable to plastic corks for many reasons,

the main one being that cork forests in Portugal and Spain support an incredible diversity of wildlife.

www.corkfacts.com.

Crisp Packets

Although crisp packets and other similar bags carry a recycling symbol, this is only relevant to the plastic offcuts from the process of making the bags in the factory. If you scrunch the packet into your hand you will see it will spring back at you, this means it is made of plastic and cannot be recycled currently.

Crockery—*see Ceramics & China*

Curtains

RECYCLE SELL*:* There is now a network of shops buying and selling old curtains—see www.thecurtainexchange.net.
DONATE*:* to charity shops.

Cutlery

SELL *or* **DONATE***:* to charity shops, jumble sales etc. Cutlery can also be taken to your Community Recycling Centre, please check with a site attendant whether it should be placed in the metal or bric-a-brac container.

Disposable Items

These are generally not recyclable, so try to avoid them—see also **Nappies**.

Disposable Single-use Cameras

REDUCE Avoid disposable single-use cameras. You can buy inexpensive fully automatic 35mm cameras, which will give better results and cost less to use than disposables. Alternatively, buy a digital camera, which allows you to only print the pictures you want or you can just look at them on your computer.

RECYCLE If you do buy a disposable camera, be sure to take it to a developer who explicitly promises to recycle the remains.

Despite the recycling claims on the boxes, less than 50% of disposable cameras are recycled.

Directories—see Telephone Directories

Dog & Cat Poo

RECYCLE This can be put in a sealed Green Cone: see under **Food: Meat & Fish**. Otherwise, wrap in paper and dispose of in dustbin or bury in your garden. As dog and cat faeces can contain parasites, keep away from children and always wash your hands after dealing with it. Do pick up your dog's poo whilst out walking – but DO NOT then throw plastic bags full of poo into the hedge or over the cliff etc. This is far worse than doing nothing. Yes, bag it – but then bin it! You can also buy biodegradable dog poo bags.

Drinks Cans—see Cans

Drinks Cartons—see Cartons

Egg boxes

REUSE They can sometimes be reused by the producer. Otherwise, compost cardboard ones or offer to outlets which sell eggs from the garden gate. Play groups, nurseries and after school clubs often welcome them for craft making—who hasn't made something from an egg box!

RECYCLE They can be recycled with other cardboard, either in your kerbside collection or at your local Community Recycling Centre.

COMPOST: They are good for the compost heap.

Electrical & Electronic Appliances

REDUCE Use wind-up or solar appliances where possible, e.g. radios and calculators (see under **Batteries**). Then you can listen to the radio during power cuts, or when you are working in the garden, without using batteries.

REUSE / *Repair:* Lots of electrical equipment is thrown away just because it is old and outdated. Local community recycling and furniture projects will advise you what to do with your unwanted equipment: much of it can be repaired and reused, and many of them employ qualified electricians to approve any repair work.

RECYCLE New legislation which came into force in January 2007, called 'WEEE' (the Waste Electrical and Electronic Equipment Directive) states that retailers must take back any electrical and electronic items bought from them. Producers will have to pay for the items to be taken apart, any reusable parts recovered, and most of the rest to be recycled. Electrical and electronic equipment contains hazardous substances which need to be taken out, reprocessed and reused. See www.netregs.gov.uk and search for WEEE.

Electricity

REDUCE reliance on the grid. Electricity comes largely from burning fossil fuels which contribute to global warming, and which we are fast running out of in the UK and are already importing. Use wind-up or solar-powered devices wherever possible. Change your energy provider to one who uses green energy—see www.energyhelpline.com. Compare suppliers and save money too!

> *Every day more solar energy falls on the Earth than the total amount of energy that the planet's 5.9 billion inhabitants would consume in 27 years.*

E

Energy-Efficient Lightbulbs

Reduce your energy use: An energy efficient light bulb can last up to 10 times longer than a normal light bulb, and if everybody bought just one energy efficient light bulb, we could shut down a power station. See www.myfootprint.org & Energy: use less, save more, published by Green Books. Also www.cat.org and www.nef.org.uk, or contact your local council for help.

RECYCLE Energy-efficient light bulbs can also be recycled at your local Community Recycling Centre, please place it in the fluorescent tubes container. Normal light bulbs cannot be recycled, due to the small amount of material they contain.

Engine Oil

It takes just 1 litre of engine oil to pollute 700,000 litres of fresh water.

RECYCLE You can take engine oil to your local recycling centre. It is illegal to pour oil down the drain or to burn it.

Envelopes

REUSE Envelopes cannot usually be recycled because of the glue (although some council collections take them), but they can easily be reused: stick on reuse labels, which are readily available, or remove the windows and compost them. However, it's a chore having to remove windows, so why not switch to totally compostable window envelopes—and tell others about them as well. General envelopes (C5 and DL) with real cellulose (which is compostable) window film are available from Paperback: ☎ 020 8980 2233. The Green Stationery Company sells envelopes with Glassine windows (also used to hold photographic negatives, and also compostable). www.greenstat.co.uk.

Fabrics—*see Clothing, Textiles*

Faxes—*see Junk Mail*

Fizzy Drink Bottles

REUSE They are useful in the garden as 'mini cloches' to protect young plants (see **Plastic** for more on this).

RECYCLE Most recycling centres now take plastic bottles for recycling, and some local authorities collect from the doorstep.

Fluorescent Light Tubes—*see also see Light Bulbs*

RECYCLE These contain toxic mercury, which can be safely recovered. One Devon-based project (see www.lamprecycle.co.uk) is now collecting fluorescent light tubes for reclamation.
Also see www.mercuryrecycling.co.uk.

Foil—*see Aluminium Foil*

Food

REDUCE We waste a third of the food we buy (6.7 million tonnes—see www.lovefoodhatewaste.com), so try to minimise food waste.

- Incorporate leftovers in other meals, and boil up bones for stock
- Most fruit and vegetables are designed by nature to keep for weeks or months in the right conditions, in the cool and out of direct light, so don't just bin it because it's a week old
- You can freeze or bottle all kinds of fruit and vegetables.
- If you grow your own, there are all kinds of ways of storing food for long periods, including drying, fermenting and bottling

REUSE
- Leftovers can be frozen and eaten another time or an easy lunch the following day (ensure you reheat any leftovers

thoroughly before eating)
- Dogs have their uses: relishing your leftovers
- Stale bread is never wasted in Italy: it can be made into croutons or used in salads, where it absorbs balsamic vinegar and olive oil dressings—yum!
- Wild birds will also eat breadcrumbs, you can make fat balls for them by mixing fat and breadcrumbs together

RECYCLE Some local authorities are now offering a kerbside collection for garden waste, kitchen waste or both.

COMPOSTING FOOD WASTE

Meat & fish

People are generally advised to avoid composting meat and fish: this is primarily because of attracting rats. However, you can use a sealed system such as:

- The E. M. Bokashi system (see www.livingsoil.co.uk)
- A wormery (for how to make one, see www.the-gardeners-calendar.co.uk, or your council may be able to help).
- A Green Cone (a proprietary sealed system which has a basket you sink into the ground and a cone above—see www.greencone.com), or
- A tumbler (see www.hdra.org.uk/factsheets/gg14.htm)
- An insulated tumbler (the Jora) – it can take anything! See www.smartsoil.co.uk
- Scotty's Hot Box—my latest project with Green Machine solutions: see www.wormresearchcentre.co.uk.

> *About 60 per cent of the average dustbin's contents are materials which can be turned into compost.*

It is best to seek professional advice before undertaking any large scale recycling of cooked foods, meat, fish, cheese etc.

Fat from cooking can be composted (mix with plenty of paper), or even better mixed with breadcrumbs and seeds and poured into half a coconut shell to feed the birds.

Fruit & vegetables
Raw fruit and veg can go into an ordinary composting system (small amounts into a wormery), or any of the other systems mentioned above.

Other cooked food (e.g. pasta & rice)
Similarly, put into a Bokashi, tumbler, Green Cone or wormery—anything that's not accessible to rats. See *Composting: An Easy Household Guide* (Green Books) for more information.

Fridges & Freezers

REDUCE your energy use by buying fridges and freezers with a good energy-efficiency rating.

REUSE Old chest freezers are also useful as rat-proof feed stores, and can be converted into composting containers or wormeries for food waste. Contact the Community Composting Network in the **Resources** section for advice. Take to a recycling centre to be reprocessed at the end of their re-use life.

RECYCLE Old fridges and freezers contain CFCs in their coolant gases and insulation material. CFCs are extremely damaging to the ozone layer, so these old appliances must be disposed of at recycling centres, where they are sent off to be shredded at specialist facilities which recover all the useful materials and harmful gases—see **CFCs**. Modern fridges and freezers do not contain CFCs. Take your fridges and freezers to your local Recycling Centre—check with your local council first where they can be accepted.

Funerals

More and more people want to leave the world without causing unnecessary pollution. You can now ask for a cardboard coffin, a woollen shroud, or even an ecopod. See www.greenburials.co.uk www.eco-pod.com www.feltcocoon.co.uk.

Furniture

RECYCLE SELL *or* **DONATE**:

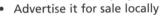

- Contact your local furniture recycling project or social services. See Furniture Reuse Network ☎ 0117 954 3571 www.frn.org.uk to find the nearest furniture project to you.
- Advertise it for sale locally
- Upholsterers want good pieces to re-cover, and for their students or apprentices to practise on
- Your local Council may arrange a 'bulky household waste' collection service.

Also see www.re-formfurniture.co.uk.

Garden Chemicals

AVOID Use safer and more effective biological controls in your garden and greenhouse instead of chemicals. Chemicals will kill the insects, which are more effective than anything at killing pests. There are plenty of alternatives to chemicals in the garden, many of which are now banned for use and a problem to dispose of safely.

If you have old stocks of garden chemicals to get rid of, contact your local council for advice on disposal.

If you want advice on gardening without chemicals, contact Garden Organic www.gardenorganic.org.uk ☎ 02476 303517. www.OrganicCatalogue.com.

Garden Tools—*see Tools*

Garden Waste
(clippings, hedge prunings, branches etc.)

Compost: Smoke from garden bonfires is 350 times more carcinogenic than tobacco smoke, and is particularly bad for children and babies. It is far better to compost garden waste *in situ* or take it to your local community project.

REUSE Clippings can be sawn up for firewood or stacked as wildlife habitat areas—a fallen tree is a wildlife refuge, as is a pile of woody prunings—so if you have space, leave a corner with a wildlife pile. Amphibians, especially newts and toads, love a pile of slowly rotting wood, as do scores of beetles, some of which are becoming very scarce because of our over-tidy gardening habits.

Don't forget to save useful sticks for plants to climb up, especially peas and beans. Woody 'brash' can also be woven into uprights to make a 'fedge' (a cross between a fence and a hedge). If you do this just in front of an existing fence, wall or hedge, then not only do you create a marvellous wildlife corridor but also a space, behind which you can pile up even more woody brash or compostable materials to slowly break down. Also see **Compost.**

RECYCLE through community composting projects (see **Resources** pages), or take it to your local Recycling Centre to be composted.

Gas Cylinders

These will normally be returnable to the supplier and can be taken to your local Recycling Centre.

Glass (sheets)

REUSE Flat glass can be cut down to make smaller panes, or made into cloches.

RECYCLE It can be taken to a household recycling centre or glass merchant for recycling. Don't try to put it into a bottle bank, because it's so dangerous to handle.

Glass, along with paper and tins, is the most common item collected for recycling. This is because it is very worthwhile to collect glass, and more and more markets are opening up for it—see below.

Glass Bottles

RECYCLE Either at the local bottle bank, through your council's kerbside collection or at your local Recycling Centre. You have to sort glass according to colour—if you have any colour other than clear, brown or green (e.g. red, yellow or blue), put it in the green section. Their lids however cannot be recycled.

What happens to glass? Besides being turned back into glass bottles and jars, glass is also used for many other applications. There are many different uses for some recyclable materials, and glass is a good example.

> *In Britain almost 200 glass bottles and jars are thrown away every second.*

- Glasspaper or sandpaper.
- Sandblasting is an industrial abrasive system, which works even better with glass, and the great thing is that it doesn't matter what colour it is. That is also true of most of the alternative uses.
- Water filtration works even better with glass than other products. Then of course there is glass fibre and fibreglass insulation. The list of alternatives is long: glass can be added to concrete (glass-crete), and to asphalt (glassphalt).
- It is used in the ceramic industry: glazes are basically a glass layer on a pot.
- It is used in hydroponic growing as a soil-less growing medium.
- Bottles are being turned into glasses: one design uses the bottle with the tapered top section cut off; another design uses the inverted top section which is stuck on to a piece cut from the bottom of the bottle.
- Bottles can be used in house construction: mortared together on their sides, or they can make translucent insulating walls.
- One entrepreneur in Plymouth has devised a way of slicing up bottles, heating them up and creating flat panes of glass

that can be used in making decorative glass panels.

- A new process can fuse together shards of different coloured broken glass and make a very solid worktop resembling granite, called Ttura: see www.eightinch.co.uk (also see the 'Urban Ore' case study in the 'Vision for the future' chapter at the end of this book).

Glasses (spectacles)

RECYCLE DONATE Take to a spectacle retailers or your local Recycling Centre for re-use in the Third World. Also see Vision Express at www.vao.org.uk/spectacles.

Grass Cuttings

RECYCLE *Compost:* Mix with scrunched up paper, card and woodchip to help aerate your mix, or leave them on the lawn. Or buy a mulching mower, which cuts the grass up very fine to incorporate back into the soil.

Greaseproof Paper

RECYCLE *Compost:* It is compostable, but does take quite a time to break down, so rip it up and scrunch it first.

Greeting Cards

REUSE Turn them into gift tags, or cut up and make new cards. Many community groups, post offices, shops, councils etc collect old cards for re-use and recycling (see also **Cards** and **Christmas**).

RECYCLE Take them to your local Recycling Centre and place them in the Mixed Paper container.

Growbags

RECYCLE *Compost:* Add to your compost heap or just sprinkle on top of your soil.

Hardcore—*see Building materials*

Herbicides—*see Chemicals*

Hire

If you only need something occasionally, why not hire it, rather than having to store, repair and eventually dispose of it?

Household Cleaners—*see Cleaning*

Hypodermics—*see Needles*

Ice Cream Containers

REUSE There are lots of possibilities for storage and freezing food.

RECYCLE if possible: check with your local council.

Inkjet Cartridges *(from computer printers)*
—*see also Toner Cartridges and Mobile Phones*

RECYCLE Cartridge World will refill your cartridges and you can save up to 60% of the cost of a new one. There are several charities who will give you Freepost plastic bags so that you can send them your cartridges, e.g. SCOPE ☎ 020 7619 7239. The World Wide Fund For Nature (WWF) recycle some cartridges ☎ 0800 435576 www.panda.org. www.officegreen.co.uk will buy inkjet and toner cartridges. Take great care of your cartridges, wrap them in the original packaging if possible, many are rejected because of damage, especially to the printed circuit.

Always turn off a printer so that it parks the print head (don't just switch it off at the mains), to prevent the print head drying up. You can also buy cartridge refill kits (but some people find these fiddly, and not always effective).

Jars—*see Bottles & Jars*

Jiffy Bags

REUSE Save for re-use.

RECYCLE *Compost:* Some jiffy bags are made from brown paper and padded with shredded newsprint, and are 100% compostable if too damaged to be reused. Others use bubble wrap, which should be removed before composting and can be reused.

Jigsaw Puzzles

REUSE DONATE: Complete jigsaws can be reused by schools, playgroups, residential homes etc, or take to charity shops, jumble sales, swap shops etc.

Junk Mail

REDUCE

- Junk Mail can be stopped by calling the Mailing Preference Service ☎ 08457 034599, or write to: Mailing Preference Service, FREEPOST 29 LON 20771, London W1E 0ZT, or by visiting www.mpsonline.org.uk
- Unaddressed junk mail can be reduced by emailing optout@royalmail.com or writing to: Freepost RRBT-ZBXB-TTTS, Royal Mail Door to Door Opt Outs, Kingsmead House, Oxpens Road, Oxford OX1 1RX
- Faxes can similarly be stopped by contacting the Fax Preference Service: www.fpsonline.org.uk or e-mail fps@dma.org.uk.

RECYCLE as paper.

Kitchen Foil—*see Aluminium Foil*

Kitchen Roll

Buy recycled: always use kitchen roll made from recycled paper.

RECYCLE Put used paper (and the cardboard centre) in the compost.

L

Kitchen Waste—*see Food*

Knitting Wool—*see Wool*

Lamps

RECYCLE Take to a recycling centre.
SELL *or* **DONATE** to charity.

Light Bulbs & Light Fittings

REDUCE Buy low energy bulbs. If every household put in just one low energy light bulb, it is estimated that we could shut down one power station. Energy-efficient light bulbs are now much better, more attractive and cheaper than they were a decade ago. Changing 24 light bulbs (a typical number for a household) could save you over £200 a year. See *Energy: use less, save more* (Green Books).

Dimmer switches help to prolong the life of conventional (incandescent) light bulbs, and dimming your lights saves energy.

RECYCLE Some groups are now recycling light bulbs, fluorescent lights, television sets and computer monitors. See www.mercuryrecycling.co.uk and www.lamprecycle.co.uk. For electric lamps (LED), which use 93% less mercury and last up to three times longer than the average bulb, see www.eurobatteries.com. Your energy-efficient light bulbs can be taken to many local Recycling Centres and placed in the Fluorescent Tube container.

Lino

RECYCLE Real lino (made from linseed oil and backed with hessian) was largely supplanted many years ago by synthetic man-made floor coverings. However, you can now get all kinds of exciting new linos made from natural ingredients, which could be shredded up and composted at the end of their useful life. See www.marmoleum.co.uk.

Magazines

REUSE DONATE: Dentists, doctors and hospitals use magazines in their waiting areas; residential homes, hospices will also use them, and they are a great source of material for children's play and art classes.

RECYCLE either through your local kerbside collection or in the newspaper bank or local recycling centre. Or they make great bedding for small animals instead of sawdust: simply line the cage with newspaper and place some shredded magazines over the top—they'll love it! (Make sure any staples are removed first.)

Material—*see Textiles*

Medicines

Return unused medicine and medicine bottles to your chemists.

Metal

RECYCLE SELL *or* **DONATE**: Take to a scrap metal dealer or recycling centre.

Milk Bottles

REUSE *Return* glass milk bottles to your milkman/supplier. These are then reused up to 13 times before they have to be recycled. Glass can be recycled again and again without ever losing its clarity or purity.

RECYCLE Plastic milk bottles can be recycled in some areas: check with your local recycling centre or council.

Mobile Phones

RECYCLE There are lots of ways to recycle your mobile:

- You can return your unwanted mobile phone handsets and accessories directly to over 1,200 retail outlets throughout the UK, e.g. O2, Orange, T-Mobile and Vodafone.

- You can collect a Freepost envelope in-store, and send your unwanted handset and accessories to the Fonebak recycling centre (Virgin Mobile, Virgin Megastores, Currys, Dixons, The Link, PC World) www.fonebak.com.
- Tesco: for every phone they recycle, £2.50 is donated to Alzheimer's Dementia Care and Research, NCH and Cystic Fibrosis Trust. Pick up an envelope in-store
- Your local Recycling Centre may also accepts them, and donate them to charity
- Actionaid, Freepost SWB131, Bristol BS1 2ZZ ☎ 0117 929 8818) for post bags to return them
- SCOPE Recycle-a-Phone, Freepost MID23462, Burton on Trent, DE14 1BR. Free collection facility for between 10 and 100 phones ☎ 0800 0832103
- Red Cross, CDM Freepost SEA10057, Leatherhead KT22 7BR will collect large numbers (over 50) ☎ 0800 0153576
- Recycling your mobile phone can help to reduce the impact that mining of the mineral coltan is having on the forests and wildlife of the Congo. Coltan is used to make the rare metal tantalum needed in mobile phones

  www.rainforestconcern.org

In the UK, the average consumer replaces his or her mobile phone every 18 months, and it is estimated that some 15 million mobile phones are replaced each year in the UK.

Monitors

RECYCLE TV screens and monitors contain mercury, and need to recycled carefully: take them to your local recycling centre. www.mercuryrecycling.co.uk and www.lamprecycle.co.uk.

Mortar

RECYCLE Reasonably clean and paint-free lime mortar and sand that comes from demolishing old brick and stone work is a good addition to acid soil. Use a length of chicken wire as a sieve to separate out pieces of rubble, then bag it up and use in the garden or allotment.

Nappies

Eight million nappies are thrown away every day, and each child uses a total of 5,850 nappies in his or her lifetime.

REDUCE Try to use disposable nappies only for when you are travelling or cannot use reusable ones.

REUSE The eight million disposable nappies sent to landfill every day in the UK will take hundreds of years to decompose. Why not try reusable nappies instead? They come in loads of different styles, are easy to use and can save you money as well! See Women's Environmental Network www.wen.org.uk. Alternatively, many areas now have a nappy laundry service.

Some councils offer cash back for each child that uses cotton nappies, if you join Cotton Bottoms Nappy Laundry Service: contact your local council to see if there are any schemes operating in your area.

Nappy Laundry Service ☎ 01798 875300 Premium Rate Line www.tommeetippee.co.uk Real Nappy Association ☎ 01983 401959 www.realnappy.com Nappyline (where to buy cloth nappies) ☎ 01983 401959.

Needles/Hypodermics

DISPOSE If you have needles or other sharp medical equipment, ring your local council, doctors' surgery or hospital to get advice on where to take them and what to take them in safely.

Newspaper

REUSE

- Newspapers can be used for all kinds of things, from lining pets' cages to protecting floors from paint.
- If you can't use it, try by donating to local nurseries and schools for use in art and crafts.
- A thick layer of newspaper is a great 'barrier' mulch, which suppresses weeds.

RECYCLE

- Most local authorities collect newspapers in their kerbside recycling collections, or you can take it to a local bring site or your local Recycling Centre.
- Your local community group may shred paper for animal bedding provided it is not contaminated with chemical spills
- Add newspaper to the compost
- Buy a flowerpot maker which transforms your newspaper into pots: see www.cat.org.uk
- Get a log maker, which turns newspaper into briquettes to burn on the fire. These can be purchased from the Centre for Alternative Technology ☎ 01654 702 400, www.cat.org.uk.

What happens to recycled newspaper? As well as being recycled back into new paper it is also turned into 'Warmcell 100', a great way to insulate your house. See www.greenshop.co.uk ☎ 01452 770629.

Recycling a one-metre stack of newspaper saves one tree.

The Office

Changing work patterns mean that more and more people are now working from home. This has the positive effect of cutting down on commuting, saving time, energy and pollution. If you work from home, you can shop around for the best recycled office products.

REDUCE, REUSE, RECYCLE

> *If each one of the UK's 10 million office workers used one fewer staple daily, there would be a saving of 328 kilos of steel a day—120 tonnes a year. Use a paper clip instead!*

REDUCE

- Use e-mail rather than paper
- Use solar-powered calculators, not battery-operated ones
- Use products with a longer life, such as low-energy light bulbs, which last up to eight times longer than ordinary light bulbs and also reduce energy costs.

REUSE

- Reuse envelopes for internal mail
- Buy envelope reuse labels, which can be purchased from charities, sometimes with your company logo overprinted
- Use re-useable items rather than disposable ones: e.g. china cups, metal cutlery, propelling pencils, refillable pens
- Packaging: see if you can have a 'take-back' option on packaging, especially polystyrene chips and bubble wrap. See www.greenlightproducts.co.uk.

RECYCLE

- Collect stamps and milk bottle tops for charity
- Turn scrap paper into notepads
- Have boxes for unwanted used paper and use it in the photocopier
- Use both sides of the paper when photocopying or producing reports
- Use recycled paper
- Find out whether there is an office paper collection in your area. This is for high quality paper, and worthwhile collecting separately
- Put used paper to one side for recycling
- Have a paper shredder for confidential documents, which can then be composted or used for animal bedding
- Save used printer/toner cartridges for reuse (see under **Mobile Phones** and **Cartridges**)

62

- Only print out documents when you need to
- Contact local community recyclers or charities if you are upgrading equipment—have it re-used!
- Office equipment and furniture: contact your local furniture-recycling project, or sell to a second hand shop
- Have an office wormery for those tea bags and fruit peelings.

Offices can be extremely wasteful places, but with a little thought and organisation they can become much more sustainable. For more support in the workplace, contact Global Action Plan: www.globalactionplan.org.uk ☎ 020 7405 5633.

> *Every year we need a forest the size of Wales*
> *to provide all the paper we use in Britain.*

Tips for purchasing for your home or work office

- Specify products with a recycled or reconditioned content. For wooden furniture and other timber products, this may include purchasing goods from certified, sustainable sources
- Avoid buying disposable products and aerosols
- Use solvent-free correction fluids and paints
- Choose local products and materials to reduce the energy and pollution implications of transporting goods
- Avoid over-packaged goods
- Buy a fountain pen, or use refillable pens and highlighters
- Consider upgrading your PCs rather than replacing them
- Share items in occasional use, e.g. hole punchers
- Order recycled products, use stapleless paper joiners
- Buy compostable pens made from cornstarch (called Mater-Bi), now with built-in seeds—when the pen runs out, plant it in the garden!
- Remarkable Pencils Ltd make pens from printers, mouse mats from car tyres, and lots more: www.remarkable.co.uk

See www.officegreen.co.uk www.greenshop.co.uk. For recycled paper suppliers www.recycled-paper.co.uk ☎ 01676 533832. General envelopes (C5 and DL sizes) with (compostable) cellulose window film are available from Paperback ☎ 020 8980 2233.

Oil—see *Cooking Oil and Engine Oil*

Organic Waste—see *Composting and Food*

Oven Cleaner—see *Chemicals*

Packaging

REDUCE Try to avoid buying over-packaged products, especially polystyrene and plastic wrapping, which is not generally recyclable. Use biodegradable starch-based chips, which you can compost! See www.greenlightproducts.co.uk.

REUSE bags, use a shopping bag or basket, and buy loose products whenever possible.

RECYCLE or compost as much as you can of the cardboard and paper packaging.

DONATE: If there is a mail order business locally, they may well want your polystyrene chips (see separate section) or bubble wrap—see **Bubble Wrap**.

Because packaging has become such a large part of our waste, various new (and some old) ideas are around to lessen the wastage.

- One Dutch group of designers is experimenting with growing gourds into specific shapes, using plywood moulds, to produce natural packaging. www.greenlightproducts.co.uk
- Foamed starch polymers, made using steam instead of harmful CFC gas, are now used frequently instead of polystyrene chips. The end product can then be composted after use.
- Another idea is the air box. Goods can be posted inside their own bubble of air inside a reusable plastic bag. Apparently IBM invited people to the launch of the product with an egg enclosed!

Packed Lunches

REDUCE PACKAGING A survey of school waste showed that there is enormous wastage, especially from packed lunches: crisps, chocolate and sweet wrappers, fizzy drinks bottles and cans. Send your child to school with refillable food and drink contain-  ers. Ask the school to get recycling facilities— your local council should be able to help you. www.alupro.co.uk have a good scheme—see also under **Schools**.

SAVE MONEY

Producing a waste-free lunch each day can make considerable savings. The table below gives you an idea of what you can save. NB These prices are merely indicative—food prices are rapidly rising at the time of this revision in 2008.

Lunch with disposable packaging	Waste-free lunch
Pre-packed sausage roll or pasty £1.50	Home-made sandwich (cheese and salad) 30p
Crisps 40p	Crisps (decanted from a larger bag) (in re-useable container) 20p
Pre-packed cake bar 40p	Slice of cake (home-made) (wrapped in foil or greaseproof paper) 20p
Drink pouch/can/bottle £1.00	Drink made from larger bottle of squash 5p
Total per day £3.30	**Total per day 75p**

Incredible savings! For further information and ideas on becoming environmentally friendly look at www.wastefreelunches.org and www.eco-schools.org.uk.

Paint

REDUCE Each year over 300 million litres of decorative paint is sold in the UK for domestic and trade use. A significant proportion of this remains unused and is eventually disposed of, often after a period of storage. Discarded paint is awkward to dispose of, and wastes valuable resources that could be used by others. Avoid toxic paints: try to use natural paints where possible.

If you purchase paints made from natural materials, plant oil extracts and simple minerals, it is far better for your health when you are painting, and any left-over paint dries up in the tin and can be composted! Natural paint in the house allows walls and wood to breathe, and natural paints smell really nice. Try real turpentine instead of turps substitute or white spirit— it is distilled from pine, and is the best air freshener I know of!

Even if you can't smell something, it doesn't mean it doesn't affect you, and strangely enough, the less the paint smells the worse it can be.

www.auroorganic.co.uk www.nutshellpaints.com
www.greenbuildingstore.co.uk.

RECYCLE DONATE: The Community RePaint scheme will find a home for your old unwanted paint so that it can be re-used for the benefit of the community. If you do not have a Community RePaint scheme locally, then community groups, playgroups, theatre groups etc may well be happy to have it. www.communityrepaint.org.uk.

DISPOSE: Some old paints contain lead and other dangerous substances and should be taken to a recycling centre for safe disposal.

Paper—*see also Newspaper*
REUSE
- Use both sides of the paper
- Make a scrap paper pad
- Buy recycled paper paper (if it says "100% post-consumer

waste" on the label, this definitely means it is 100% recycled from used paper.)

RECYCLE Each tonne of paper we recycle saves 17 trees being cut down to make new paper. Managing our insatiable demand for timber should reduce the need to clear old growth forests which are rich in biodiversity.

COMPOST soiled and crumpled paper.

Less than half of the paper used in the UK is recovered and over five million tonnes gets dumped in landfill sites every year.

Peat

REDUCE / AVOID Our precious peat bogs are being destroyed by peat extraction. Peat bogs take thousands of years to form, and are a unique ecosystem. At the same time as we are destroying peat bogs, we are landfilling or burning materials (especially from our gardens) which we can compost to make a wonderful peat substitute.

REUSE If you have any peat from bought-in plants it can be added to the compost heap or just spread in the garden.

Pesticides—*see Garden Chemicals*

Photographic Chemicals

DONATE: These contain silver, which is worth reclaiming. See www.igp-ukltd.co.uk.

Pillows—*see Bedding*

Plastics

REDUCE It may be impossible to avoid plastics, but we can at least try to minimise the amount of plastic we use and throw away by:

- Refusing plastic bags
- Refilling containers
- Purchasing alternative products
- Buying products in glass bottles where possible

RECYCLE Plastics will undoubtedly be more widely recycled when more facilities for dealing with mixed plastics are built. Some local authorities in England collect mixed plastics, including plastic film but excluding expanded polystyrene. Councils which collect for recycling concentrate on plastic bottles, because they are made of the highest value plastic. Your local council recycling centre should have a plastic bottle collection skip, so that the most dedicated recycler can save up plastic bottles to drop off when passing.

New Uses for Recycled Plastic

There are masses of applications for recycled plastic in contemporary design products—including furniture, clothing, lamps, screens, construction, flooring etc., either as single plastics or even as mixed plastics fused together.

Designers are also coming up with new uses for plastic objects: for instance a plastic fishing float turned into a light, and plastic bottles linked together to make a floating lounger for use in the swimming pool. Another design uses a group of plastic milk bottles with lights in and it makes a surprisingly elegant light with subtle diffusion. See the *Eco Design Handbook* by Alastair Fuad-Luke for lots of design ideas.

3,300 million plastic bottles were recycled in 2007.

Plastic Bottles

The majority of plastic bottles are made from PET or HDPE—the most valuable and worthwhile plastic to recycle.

RECYCLE Make sure you wash (and preferably squash) your bottles and remove the lids; some councils now accept plastic bottles

in their kerbside recycling collections. Alternatively, some councils now have plastic bottle recycling banks, and your local Recycling Centre should also accept them. Plastic bottles can be made into fleeces (amongst other things): see www.patagonia.com.

REUSE

- Garden Organic sell special 'dripping taps' which you can screw into a plastic bottle to water plants, useful when you are away, etc. See www.OrganicCatalogue.com, *enquiries@chaseorganics.co.uk* ☎ 0845 130 1304.
- Also use as a mini cloche to protect plants.
- You can utilise the whole bottle to make a self-watering system. Cut the bottle in half and fill the bottom half with water. The top half you turn upside down and fill with compost and your plant. Put the top half into the bottom. You can then regulate the water by loosening or tightening the top.

> *Recycling a single plastic bottle can conserve enough energy to light a 60W lightbulb for up to 6 hours.*

Plastic Carrier Bags

Plastic bags are one of the worst litter hazards in the world – see www.mcsuk.org. They easily blow into the air and get washed down rivers into the sea, where they kill marine life. Turtles eat them, thinking they are jellyfish. Rebecca Hoskins, the BBC camerawoman who made Modbury in Devon the first plastic bag-free town, has a most useful website: www.plasticbagfree.com. She also made a most inspiring film called 'Message in the waves' see www.messageinthewaves.com.

REDUCE / *Refuse* bags:

- Take your own! Keep a string Turtlebag www.turtlebags.co.uk in your pocket, or have an Onyabag www.onyabags.co.uk on your keyring.

REDUCE, REUSE, RECYCLE

- Say 'no' to plastic bags, and 'yes' to jute and cotton—see Rebecca's site and www.canby.co.uk
- Some supermarkets now have 'degradable' bags (made from petrochemicals) which go brittle after a while and fall apart. But microscopic plastic is hoovered up by filter-feeding marine animals, and gets into the food chain. A better solution is the 'compostable' Bio bag: www.biobag.no.

Since the Irish government put a tax on plastic bags, there has been a 95% reduction in bag use, and the problems with them littering the streets disappeared overnight.

> *We use eight billion plastic bags each year: more than 300 for every household.*

REUSE
Get over the plastic bag habit, but re-use up any you have rather than landfilling them, e.g.:

- Scrunch them up and use instead of bubble wrap
- For packed lunches: use paper bags or bio bags in future
- For freezing: bio bags are just as good
- Reuse old plastic bags don't buy any more rolls of new ones

RECYCLE Use in-store recycling bins, or offer them to shops that don't have their own bags e.g. charity shops, or any other local store may be interested.

> *80% of our plastic waste ends up in landfill.*

Plastic cups
REDUCE If you are having a party, you can usually borrow glasses for free from your local wine shop, as long as you buy

some wine from them. And you only pay if you break any.

RECYCLE Used plastic cups from vending machines can be recycled into plastic rulers and pencils. So if you work in an office with a vending machine, contact one of these companies: www.remarkable.co.uk and www.save-a-cup.co.uk.

Plastic trays (around vegetables, fruit, etc.)

REDUCE Avoid where possible—try and buy fruit and vegetables without this packaging. You could resort to unwrapping it at the checkout and leaving it for the store to dispose of, as shoppers here and in Germany are starting to do.

REUSE as seed trays etc.

Plastic yoghurt pots

Thick or thin? Some manufacturers have made their pots as thin as possible to minimise the plastic used, and strengthened them with an outer wrapping of cardboard, which can be removed for composting or recycling.

A few manufactures have taken the opposite approach and have put their yoghurt in sturdy reusable pots. However, there are only so many you can use!

REUSE as seed pots, paint pots, storage containers or for children's playgroups / after school activities.

Plastic Packaging

Over 60% of the total plastic waste in Western Europe comes from packaging, which is typically thrown away within one year of sale.

Hard polystyrene preformed packaging (e.g. used to protect electrical appliances)

DISPOSE Try to get the suppliers to take it back; otherwise you will have to put it out for the rubbish collection.

REDUCE, REUSE, RECYCLE

What happens to polystyrene?

Polystyrene can be made into new products including a replacement for hardwood, which is suitable for garden furniture, window frames and picture frames.

Expanded Polystyrene (EPS)

This is the soft cushioning plastic which many goods come wrapped in, e.g. fridges, televisions, computers and goods sent through the post. It is recyclable, although it is really only feasible for the trade to recycle it. In theory, the supplier of the goods should take back any packaging, and we should all be insisting that they do just that.

REDUCE There is a compostable version called Eco-Foam made from corn starch (or use air-popped popcorn – without oil). See also **Biodegradable plastics** below.

REUSE If you know anyone who has a mail order business, or runs a shop where they need to pack fragile items, they may be happy to use this kind of packaging.

RECYCLE To recycle polythene packaging, carrier bags and wraps from magazines, cut out any paper labels (which clog the recycling machines), enclose your name and address and send to Polyprint (see below). If you send items that they are unable to recycle, they will send them back to you for future reference. The company only recycles high density (HDPE) or low-density polythene (LDPE), i.e. recycling polymer symbols No. 2 or No. 4.

How can you tell if it is recyclable? If there is no symbol on the plastic, try stretching it. If it stretches it is likely to be polythene and can be recycled. If it snaps it is probably cellophane or PVC, which cannot be recycled by this company.

PolyPrint Mailing Films Ltd, Mackintosh Road, Rackheath Estate, Rackheath, Norwich NR13 6LJ ☎ 01603 721807. See www.polyprint.co.uk/recycling.html.

With so many different types of plastic used in packaging, each with its own symbol, is it any wonder that people get confused by what they can and cannot recycle? There are about 50 different types of plastic. www.recoup.org gives information on plastic and plastic products. Alternatively, a number of big chain supermarkets accept plastic bags for recycling in their store.

Frequently Asked Questions

What do the codes mean?

The most common forms of plastic packaging found in the home are often marked with a code. This helps us to know which plastic is which. The code can be either a number or letters, and is usually found in or with a recycling sign, but you may still have difficulty in finding somewhere to recycle plastic near your home.

What do the numbers mean?

⚠ 1 or PET = Polyethylene terephthalate
    (e.g. fizzy drinks bottles and water bottles)

⚠ 2 or HDPE = High density polyethylene

⚠ 3 or PVC = Polyvinyl chloride (e.g. food trays, detergent bottles, food wrap, vegetable oil bottles, blister packaging)

⚠ 4 or LDPE = Low density polyethylene (e.g. plastic bags, bin liners, sandwich/bread bags, six-pack rings)

⚠ 5 or PP = Polypropylene (e.g. margarine tubs, straws, refrigerated containers, screw-on bottle tops/lids, some carpets, some food wrap)

⚠ 6 or PS = Polystyrene (e.g. yoghurt pots, styrofoam cups, throwaway utensils, meat packing, packaging chips)

⚠ 7 other = Polycarbonate, acrylic, ABS, mixed / multi-layer plastic.

Numbers 1 and 2 are the most commonly recycled plastics at the moment. Hopefully in future we will be able to recycle them all.

What is plastic made from?

Plastics are made from oil—usually mineral oil (petroleum), although plastics made from plant material are becoming more common.

Are there any alternatives?

Biodegradable Plastics: Plastics made from plant materials like corn starch are useful as they are fully bio-degradable and can be composted: see *Green Plastics: An Introduction to the New Science of Biodegradable Plastics* by E.S. Stevens.

- These bags can be used for kitchen waste to keep the bin clean, and to contain kitchen waste for compost collection rounds
- They are also used as dog poo bags

Bio-plastics are also used for compostable plates and cutlery. The maize plants at the Eden Project are being mulched with maize cutlery! In Australia, kangaroos and other animals soon eat biodegradable fast food boxes tossed out of car windows, or they bio-degrade naturally—perhaps not the most elegant of solutions, but at least it cuts down on plastic litter. www.biopolymer.net.

> *25 recycled PET bottles can be used*
> *to make an adult's fleece jacket*

Is plastic recycling really worthwhile?

Much of our plastic finds its way, ultimately, to the sea, where it degrades into smaller an smaller pieces and is eaten by a variety of life forms. Eventually the smallest fragments are ingested by plankton and filter feeding fish and other animals and thus accumulated through the food chain. We have a real need to stop wasting and throwing away plastics so we simply must develop better 'closed loop' plastic recycling.

Genuine recycling, where there is no loss of quality and materials can be endlessly 'upcycled' into similar or better products, is viable. However, where plastics are mixed they can only be 'downcycled'; this should really only be seen as a short term solution as everything produced should be either compostable at the end of its useful life or endlessly recyclable without loss of quality. A further problem is people mixing in biodegradable plastics into plastics for recycling when they should be composted! Having said that, recycling plastic can be cost-effective. Councils currently spend millions every year on the collection and disposal of plastics. Through effective collection schemes they can sell plastics and generate income instead. Usually when plastic is recycled it is changed into another type of item. For example fizzy drink bottles (PET bottles) can be recycled into insulating material for synthetic fleece jackets, or back into recycled bottles.

There are some plastics factories which can take mixed plastics. At the Intruplas factory at Sowerby Bridge, Yorkshire, plastics are shredded and mixed together, then heated. The plastics with the lowest melting temperature fuse around the unmelted plastics, and the whole lot is extruded to form a variety of chunky shapes, bollards, planks, slabs etc. Shredded tyres can be added to give a non-slip surface, and woodchips can be added which helps make the final product lighter. The end products have a number of uses—canal bank strengthening, non-slip bicycle paths, pub garden furniture, civic planters— and more uses are being found all the time. See www.greenbusinessnetwork.org.uk (type in Intruplas).

Postcards

REUSE DONATE: Old postcards, used or unused, can be sent to Actionaid, where they are sold to collectors to raise funds. See www.impact-initiatives.org.uk.

RECYCLE Postcards can be recycled in the Mixed Paper container at your local Recycling Centre.

Prams, Pushchairs & Cots

RECYCLE SELL, DONATE, FREECYCLE *or* **SWAP IT** Advertise locally contact your local social services or advertise on the Freecycle or Swap It web pages,

Printer Cartridges—*see also Ink Jet Cartridges & Toner Cartridges*

REUSE Why not refill your printer cartridges? It can also save you money!

RECYCLE DONATE: Many charities now collect printer cartridges including ActionAid, Freepost SWB131, Bristol BS1 2ZZ. www.actionaidrecycling.org.uk. Treat them carefully: try to wrap in original packaging.

Quilts—*see Clothing*

Radiators

RECYCLE Take to a recycling centre.

Radios—*see Electrical Appliances*

Rags

REUSE for cleaning, paint or car rags, or can be recycled for papermaking etc—see **Textiles**.

Rechargeable Batteries—*see Batteries*

Record Players—*see Electrical Appliances*

Use your granny's wind-up—great for picnics!

Records, Tapes

RECYCLE SELL *or* **DONATE**: to charity shops. Record collectors may also be interested in rare records and tapes in good condition.

Refrigerators & Freezers—*see Fridges & Freezers*

Sanitary protection

Never flush down the toilet. Choose organic unbleached tampons and sanitary towels with minimum packaging. Use alternative sanitary protection—contact the Women's Environmental Network in the **Resources** section for more information. www.wen.org.uk.

Sawdust

REUSE Add in small quantities to compost heaps—great to have in a container by the heap to mix with wet materials. Sawdust should only be reused in hamster, gerbil, rat or mice cages. Rabbits and guinea pigs all can develop breathing problems from being kept on sawdust, so use newspaper to line their cages, with shredded magazines or other paper on top. It makes a great toy as well!

Schools—*also see 'Children'*

Where better to start putting across the three Rs (Reduce, Re-use and Recycle of course!) than at school!

There are all kinds of initiatives, from aluminium can recycling for school fundraising to whole school eco-makeovers. Many local authorities are promoting the three Rs either by going into schools themselves, by sponsoring theatre groups or specialists to work with the children, or by doing a 'waste audit'. During a waste audit, all the waste in the school is collected up for a week and then emptied out on a huge piece of tarpaulin spread out in the gym or playing field. Everything is sorted into categories, and the children can see how much wastage is going on and start to think about ways to reduce it.

Children love to recycle and compost—is your local school doing its bit? www.alufoil.co.uk www.eco-school.org.uk www.globalactionplan.org.uk. Oxfam's education site is www.oxfam.org.uk/education.

WWF produce a huge range of publications on all aspects of sustainable development. 'Making it happen: Agenda 21 and Schools', which has case studies from UK schools, is particularly useful for inspiration and classroom activities. www.wwf-uk.org.

Scrapstores

There are approximately 80 scrapstores in the UK. They are more formally known as 'resource centres'. Find out if you have one locally: they are a veritable treasure trove for schools, playgroups, theatre groups, artists etc. They will collect all kinds of pre- and post-industrial offcuts from factories—wood, plastic, fabric and paper, including ends of rolls etc. They will also take part-used tins of paint, glue, solvents etc., providing they are in sufficient quantity and quality.

You also get all kinds of weird and wonderful 'one-offs', such as a whole container filled with plastic boxes that once held stick-on letters for signs, a load of redundant marketing videos or CDs, old shop dummies and strange display equipment—go and see for yourself! www.childrensscrapstore.co.uk.

Sharps—*see Needles*

Shoes

REUSE Charity shops and some clothing banks accept shoes in good condition—in pairs!—either tied together or in a plastic bag. There are various textiles banks throughout the county or you can find one at your local Recycling Centre.

Soap

RECYCLE You can buy soap presses which make new bars, or you can press them together yourself, just run them under the tap to make them soft and press them together. You can also chop them up to make a liquid soap.

Spectacles—*see Glasses*

Stamps

RECYCLE Compost or **DONATE** to charity shops, e.g. www.oxfam.org.uk/shops & Royal National Institute for the Blind. www.rnib.org.uk > fundraising > recycling.
SELL to collectors or specialist shops

Telephone Directories

REUSE Great for raising the height of your computer monitor!
RECYCLE

- Tear them in half and add to your compost bin. (That's what I do. OK—I do it a page at a time!)
- Shred up for animal bedding. You may have a community project or small business near you which will do this.
- White telephone directories can go into magazine/newspaper recycling banks, or be collected through your kerbside recycling scheme.
- Yellow Pages can be taken to your nearest Recycling Centre and placed into the cardboard skip. In some areas Yellow Pages can be placed into kerbside recycling boxes/bins (check with your local authority).
- Recycle through the Yellow Pages recycling project.
- Some local authorities accept Yellow Pages in their kerbside scheme—check with your local council. www.yellgroup.com hotline ☎ 0800 671 444 or ypdirectoryrecycling@yellgroup.com

Textiles—*see also Clothing, Curtains*

REUSE SELL *or* **DONATE**: Charity shops, jumble sales, or you can sell them yourself.

RECYCLE SELL *or* **DONATE**: put in clothing banks.

What happens to old textiles?

Clothes from textile banks are sorted and the clothes that are reusable clothes are sent to developing countries. Those which cannot be reused are used as rags in industry or as stuffing for sofas, mattresses etc.

There is an increasing number of entrepreneurial businesses that take used clothing and reinvent them as new 'one-off' pieces, or utilise the fabric for new designs altogether. This is an extension of the philosophy of 'make do and mend', long practised by skilled needlewomen with the design flair to create a natty new outfit from a bag of jumble sale leftovers. For many who sell their handiwork through craft shops, it is a money-earning hobby; for a very few it may provide a livelihood. See for example www.fabrications1.co.uk www.hiddenart.com. www.traid.org.

Tiles

RECYCLE SELL Good clean tiles in reasonable quantities can be sold: advertise locally or take to an architectural reclamation yard or council recycling centre. Interesting old tiles can also be sold through small ads, car boot sales etc.

Timber

REUSE wherever possible—hang on to it, as you never know when it may come in useful in the future!

RECYCLE Offcuts can be used for woodburning stoves. Tanalised wood, blockboard, plywood, MDF and chipboard contain toxic substances, and should not be burnt. However, most types of wood can be recycled back into chipboard so take to your local recycling centre.

SELL or **DONATE** Good quality wood is always useful, and can be re-sold. Some community schemes now have wood re-use facilities. Scrapstores will also take good quality clean timber. See also www.recycle-it.org.

The amount of wood presently wasted in the UK is about seven million tonnes per annum. This is made up of wood from the construction, demolition, joinery and wood products manufacturing industries, wood packaging waste and pallets, green wastes and a substantial amount from the domestic waste stream. Only around ten per cent of all timber waste is recycled. Most of this goes to be remanufactured into chipboard and MDF.

Tin Cans—see Cans
Tin cans are in fact made mostly of steel.

Toner Cartridges—see also Inkjet Cartridges

> *In the UK over two million non-biodegradable toner cartridges are thrown away annually.*

RECYCLE
- ActionAid and Oxfam will collect some toner cartridges for recycling, but not all models. Some can be returned direct to the manufacturer (e.g. Hewlett-Packard, who will supply you with boxes for them and then arrange collection).
- Collect them in quantity—have a collection point at work or at school. Once you have collected 10 or more, contact one of the charities below (they often collect mobile phones as well).

ActionAid Recycling Greensource,14 Kingsland Trading Estate, St Philips Road, Bristol BS2 0JZ ☎ 0117 304 2390. www.actionaidrecycling.org.uk. www.officegreen.co.uk will buy toner and inkjet cartridges. See also www.oxfam.org.uk.

Tools

RECYCLE SELL You can often sell tools successfully at car boot sales.

DONATE: 'Tools For Self-Reliance' (TFSR) is a charity that refurbishes tools to be sent abroad—they have groups of volunteers all over the country.

At the heart of TFSR is a simple idea—you can't work without tools. They are able to deliver effective targeted aid to groups of artisans in poor communities, and ensure that the right tools get into the right hands.

Tools for Self Reliance, Netley Marsh, Southampton SO40 7GY ☎ 02380 869697 www.tfsr.org.

Toys and Games

REDUCE Use your local toy library if you have one.

REUSE SELL *or* **DONATE** to friends, playgroups, residential homes, hospitals, charity shops, church groups, jumble sales etc.

Tyres

REDUCE Carry out regular maintenance checks, such as tyre pressure, to extend the tyres' life, and you may also find it benefits your fuel consumption too.

REUSE Scrap tyres make excellent compost bins or wormeries (place several tyres one above the other), plant holders when filled with soil (very good for potatoes!), children's swings and play equipment, boat fenders and even houses—see *www.earthship.co.uk*.

RECYCLE The metal rim, once it has been separated from the tyre, can be recycled at Recycling Centres. The Remarkable Company make mouse mats and pencil cases from old tyres. Most garages will take unwanted tyres for recycling or safe disposal. Unwanted tyres must not be placed in your household rubbish. www.remarkable.co.uk.

What happens to old tyres?

They can be ground up to make playground surfaces and mixed with some plastics—see **Plastics**. There is even a method for adding micro-organisms (they found sulphur-eating bacteria in the hot springs in Yellowstone Park), which eat up the sulphur used in the vulcanisation process. This makes the rest of the tyre easier to recycle.

Vending Cups

REUSE Wash out and reuse for parties, or put a hole in the bottom and use for growing seedlings and small plants.

RECYCLE The Remarkable Company makes pens and pencils from vending cups.

The Save A Cup recycling scheme collects four million of these every week from 2,000 member companies. www.remarkable.co.uk and www.saveacup.co.uk.

Britain throws away 4 billion plastic cups every year.

Water

REDUCE It's strange that some people still have an aversion to recycling when our most basic commodity, water, is being constantly recycled all the time. New housing constantly increases the demand for water, and for that reason a range of water saving measures are often included in newly built houses. Whatever the age of your house, you can do your bit to conserve water:

- Have a shower rather than a bath (but power showers can use as much water as having a bath)
- Turn off the tap when brushing your teeth
- Don't wash up under running water
- Install taps which act like showers, instead of delivering a solid stream—see tapmagic www.tapmagic.co.uk
- Install dual flush toilets
- Put a brick (or a hippo! – see www.hippo-the-water-saver.co.uk) in your toilet cistern so that it uses less water
- Water your garden at night so the water doesn't evaporate but soaks in
- Get water butts and save rainwater (water companies usually

offer cut-price water butts, and also promote composting)
- Fit devices to down pipes and gutters to capture water—better for plants than chlorinated tap water. See see www.organiccatalog.com/catalog
- Compost—a garden with plenty of humus in the soil from adding compost holds rainfall better and needs less watering

RECYCLE
- You can divert your sink and bath water to use in the garden.

Water Filters

RECYCLE Brita Water Filters: recycle them by collecting up six and sending them to: Brita Recycling, FREEPOST NAT 17876 Bicester OX26 4BR ☎ 0844 742 4800.

Wellies

REUSE Cut down wellies to use as slip-ons. Plant things in them.

RECYCLE DONATE: to charity shops—in pairs! Old PVC wellies can be recycled by Dunlop: Dunlop Welly Recycling Campaign ☎ 01606 592041.

Wood—*see Timber*

Woody Prunings

Pile up in a corner of your garden to slowly rot down. In the meantime it will be a refuge for amphibians and beetles—great for keeping the pests down.

Wool

RECYCLE *Compost:* if moth-eaten or too old to reuse.

DONATE to scrapstores, residential homes, charity shops.

Workplace

- Take ideas from this book into your place of work
- Talk to your manager or boss about putting into practice some money-saving measures (you're sure to get a favourable reaction if you put it like that!)
- Suggest that everyone is asked for their ideas about how to reduce waste
- Have boxes for paper only used on one side for photocopying less important documents, etc
- Have a box for collecting ink cartridges
- Get a wormery or food digester (e.g. green cone) for packed lunch waste, tea bags, coffee grounds etc
- Look into buying products made from recycled materials
- Volunteer to do some research
- Volunteer to collect up certain items to take to the local recycling centre

Writing paper

REUSE Always buy recycled paper, and use both sides. If the label reads '100% post-consumer waste' it is definitely 100 per cent recycled. 'Environmentally friendly' can mean a whole host of things, but necessarily that the paper is recycled.

RECYCLE / *Compost* after use. Alternatively your local Recycling Centre may have a Mixed Paper container.

Yellow Pages—*see Telephone Directories*

Yoghurt Pots—*see Plastics*

Zero Waste

What's left over! You can compost your hair shirt now, and use your dustbin as a wormery! Now that you've read this guide, hopefully you are well on the way to achieving that empty bin! More ideas on www.recyclethis.co.uk and www.zerowaste.org.

Chapter 4

What Can I Do?

There's lots you can do to make a difference:

Easy

Lighten your dustbin—take out everything that could be composted and use your current recycling service. Don't use that plastic box to store toys or tools: it's for recycling! Most of us now have some sort of doorstep scheme. Read the information that comes with it, or contact your local council for more information. If you have internet access, look on your local council's website.

- Avoid over-packaging. Do you need to buy items individually wrapped? If so, is the wrapping material made from recycled materials? Could you buy the same product made from recycled materials?
- Take a bag when you go shopping—put plastic carrier bags in your pockets so you always have some handy
- Contact your council or recycling group before you clear out the garage or loft, and see what you can take (and where) for reuse or recycling. See **Bulky Refuse** in the A-Z list, or individual headings
- Use your consumer power in the workplace too. For example, nearly every office has a photocopier. Does it use recycled paper? Has it ever been tried? Do you collect paper only used on one side for re-use? Tell your boss how the business can save money by recycling
- Choose longer-life, energy efficient, solar powered and rechargeable products
- Try out a low energy light bulb

- Give up burning rubbish on bonfires
- Buy recycled products, e.g. recycled toilet paper, kitchen rolls, tissues, refuse sacks, writing paper and envelopes. Switch to recycled paper in your printer
- Use proper cups, plates and cutlery rather than plastic or paper disposable items
- Keep asking questions at your local store about the availability of recycled products. Congratulate the store manager when new recycled content product ranges are stocked
- Shop at charity shops
- Help spread the Reduce, Reuse, Recycle message (the three Rs)

A bit more effort

Can anyone else use your cast-offs? Try your local playgroup, school, charities, community group, community hall or social services.

- Compost all your garden and kitchen waste—see the **Compost** section for details on this.
- Collect plastic bottles to take to recycling centre.
- Look at what you are currently throwing away (e.g. plastic bottles, aluminium foil, clothes, boots and shoes, etc), and see if you can find places locally to take them: see the A-Z guide and contact your local council and, or recycling group for more information and help
- Buy remanufactured printer cartridges for home or workplace computers, and return spent cartridges for recycling. (see A-Z list)
- Buy loose fruit and vegetables—refuse excess packaging
- Source local produce. Use your local shops and services
- Plan any building or renovation project with recycling in mind—advertise materials you will have in advance. Educate your builders. Remember that you can sell metals—lead and copper are especially valuable. Cables contain the highest grade copper

- Buy reclaimed materials for building projects. Source fixtures and fittings from salvage yards
- Furniture and household goods can go to people on low incomes: contact Furniture Recycling Network or Social Services (see **Resources)**
- Switch all your light bulbs to low energy ones
- Buy a paper shredder and use the shreddings for pet bedding. Then you can compost soiled bedding (with your kitchen waste, naturally)
- Search the web for recycled goods—check out the websites in the A-Z guide as a starting point.

In Europe 3,500 bins of rubbish are thrown away every minute.

Go the whole hog!

Volunteer to help your local recycling/composting/furniture project—contact the Community Recycling Network, the Community Composting Network or the Furniture Reuse Network for more details on a scheme near you (see **Resources**):

- Find out even more about waste minimisation, composting, re-use and recycling.
- Find out if you could be involved in going into schools to spread the message—contact Global Action Plan, your local council, eco-schools projects, etc.
- Go to council meetings and ask why they are not doing even more.
- Become a collector of those 'fringe' recyclable items which charities collect, such as corks, metal foil, printer cartridges, jumble, books, bric-a-brac etc.
- Start your own resource centre or recycling/composting project—see **Resources** section for Community Composting Network, Furniture Reuse Network, Community Recycling Network, etc.

What can I do?

- You could take part in a Master Composter training programme: contact Garden Organic for more information ☎ 024 7630 8202 www.compost-uk.org.uk.
- You could help your local council to promote composting in the community: contact your local recycling officer and see if there are any opportunities available locally. The Community Composting Network could also help you here: ☎ 0114 2580 483 www.communitycomposting.org.
- Be a 'positive' shopper – try and only buy things that are grown locally, produced locally, or fair traded and or organic. Think every time you buy something. What's it made of? Can it be composted, or reused or recycled at the end of its life? Has it harmed anyone or the environment during its production? Should we be making anything that has a negative impact on the planet? Consider this quote from *Cradle to Cradle* by William McDonough & Michael Braungart:

"All the ants on the planet, taken together, have a biomass greater than that of humans. Yet their productiveness nourishes plants, animals and soil. Human industry has been in full swing for little over a century, yet it has brought a decline in almost every ecosystem on the planet. Nature doesn't have a design problem. People do." (See www.mbdc.com)

We'll have a look at this some more in the next chapter.

Chapter 5

A vision for the future

In the future, people may not be talking about 'waste': resources will be so valuable that we will conserve them much more carefully.

Zero Waste

'Zero waste' is also a concept catching on in many places. New Zealand and Australia already have whole districts and states declaring themselves advocates of zero waste. In the UK, Wales has declared itself a zero waste zone, and Chew Magna near Bristol is a zero waste village: see www.gozero.org.uk.

Zero Waste is a philosophy and a design principle for the 21st century. It includes 'recycling', but goes beyond this by taking a 'whole system' approach to the vast flow of resources and waste through human society. Zero Waste maximises recycling, minimises waste, reduces consumption and ensures that products are made to be reused, repaired or recycled back into nature or the marketplace.' www.zerowaste.com www.grrn.org.

A new way of thinking

If things can't be recycled, repaired, reused or composted, then should they be produced in the first place? Already there are plenty of examples of substituting products with those made from recyclable materials: e.g. the ubiquitous fast food polystyrene box can now be made out of natural biodegradable material. In Australia the kangaroos eat the packaging! I've seen a frisbee in Brighton made from hemp oil.

As toxic materials like PVC become outlawed they will eventually cease to be in the 'waste stream', which will then of course become a 'clean resource stream'. The ultimate goal is for the cycles to mimic natural cycles as far as possible, rather than having broken cycles (as we do now), which mean more and more exploitation of our environment.

Inspiration from Abroad

A resource park: **Urban Ore in California**

Instead of having waste tips, we should be developing resource parks. This is what Dan Knapp at 'Urban Ore Inc' in Berkeley, California has done. Dan spent a happy period of his life as a scavenger at a landfill site. Whilst there, he realised that all the discards could be put into just twelve categories: paper, metal, glass, plastic, textiles, chemical, putrescible, wood, ceramic, soils, plants, and reusable.

The Urban Ore operation collects source-separated reusable items, recyclables, organics and residuals. Reusable items are taken for refurbishment or repair, and are sold on site. This part of the operation includes a training programme. The recyclable materials are sorted, baled up etc for sale. Some of these are reprocessed on site. For instance, a local inventor has figured out a way of fusing together mixed crushed glass, which is made into kitchen worktops and resembles granite. These are sold at over $100 per square foot. (Also see www.eightinch.co.uk.)

There are also artists working on site retrieving interesting items to be made or incorporated into artwork. Wood for reuse or for chipping is taken next, and then all plant material and putrescibles for composting. Chemical wastes are another stage of the operation, sorted into their chemical families wherever possible. The residuals are then sorted and picked over. Any items which can be salvaged at this stage are taken out, e.g. cardboard, paper, cans etc. Finally, this residual stream is also composted in a separate operation. When it's stable and sanitised, it's landfilled. The landfill site is different from any other. It has no smell, and no flocks of seagulls, vultures or crows. Obviously in time it's hoped that the residual part of the operation will dwindle to as near zero as possible.

REDUCE, REUSE, RECYCLE

Imagine having 'resource centres' like this, where you could take the family for a good day out! They are being planned here right now, with educational facilities, small innovative business enterprises, and linked into the local food chain that will sell local food products and serve local food in the café. Then you could 'shop till you drop' with a clean conscience.

San Diego County, the home of Urban Ore, won't collect any rubbish that is deemed recyclable!

> In the Netherlands they have kerbside shoe collections, and old computers are returned to manufacturers.

A town: Curitiba in Brazil

Curitiba was a very fast-growing city in the early 1960s, and a group of architects managed to persuade the mayor that the city needed some careful planning. A competition was organised, and the mayor, Jaime Lerner, started to put his visionary ideas into action. One of these ideas was to minimise waste.

Curitiba's citizens now separate their rubbish into just two categories—organic and inorganic—for pick-up by two kinds of vehicles. Poor families in squatter settlements that are unreachable by trucks bring their rubbish bags to local centres, where they can exchange them for bus tickets or for eggs, milk, oranges and potatoes, all bought from outlying farms.

The rubbish goes to a plant (itself built of recycled materials) that employs people to separate bottles from cans, and from plastics. The workers are handicapped people, recent immigrants and alcoholics.

Recovered materials are sold to local industries. Styrofoam is shredded to stuff quilting for the poor. The recycling programme costs no more than the old landfill, but the city is cleaner, there are more jobs, farmers are supported, and the poor get food and transportation. Curitiba recycles two-thirds of its garbage—one of the highest rates of any city, north or south.

A vision for the future

For more information about this truly amazing city, see www.dismantle.org/curitiba.htm.

In researching this book, it has been fascinating to see all the small entrepreneurs being innovative with the materials that we discard. Much of the inspiration for this has come from the so called 'third world', where of course virtually anything thrown away is seized upon and used to make something else: see www.artworksforafrica.com.

- Toy vehicles in Africa made from coat hangers and old tins
- Sandals with tyre tread soles, even hammocks made from tyres
- Plastic containers of all sorts used for water containers (being lighter than the traditional ceramic)
- Earrings made from old typewriter keys (and flip-flops!)
- Beaded anti-fly door screen made of strings of beads from washed up flip-flops
- Cufflinks from Scrabble pieces
- Clothing from old photo negatives and plastic bags, even old CDs
- Baskets woven from telephone cable and old plastic bags— very colourful
- Bags from metro fabric, carpets and tyres
- Whole houses from tyres, glass bottles and tin cans

And these are some incredible art works from the discards of our society—worthy of a lavishly illustrated book!

> *In Switzerland, tax from bottle manufacturers is paid to communities that recycle.*

An individual: **The Man Who Throws Nothing Away**
There's a man in Washington State called Van Calvez who has not thrown anything away for 15 years. Well, hardly any-thing— now and again he has to do it, if only to stop drawing

attention to himself! What started as a hobby has now become his career. By carefully thinking about what he is buying in the first place, and by recycling and re-use he has managed over fifteen years to reduce his waste stream so much that it fits into just four boxes in his home. This forms a part of the display materials that he travels around with. Despite the fact that he stockpiles stuff the rest of us would hurry to hurl, he's not what the psychiatric field would call a hoarder. 'I think our society is addictive and compulsive anyway. So if I'm doing something obsessive, that's good, that's OK with me.'

'Garbage is a relatively new concept,' he says. 'A hundred years ago there was very little waste at all.'

Van Calvez maybe a crazy visionary, but even his wife has started hoarding the polystyrene clam shell packaging that she cannot bring herself to throw away. I know how she feels: I've got two bags of it myself! See www.sustainablebainbridge.net.

Over to you

This book doesn't claim to have all the answers—maybe you can fill in some of the gaps?! If it makes you pause for thought and take a look at what you throw away, like Van Calvez did (he tipped the whole dustbin out on his kitchen floor and categorised everything!), then it has achieved its purpose.

If we all stop every now and then to wonder where all these things come from and where they are going, maybe we will start to make better choices about what we buy and what we do with it at the end of its useful life (for us). We can stop, or at least diminish, the gigantic flow of discarded materials that go into holes in the ground or up in smoke. We can have fun doing it, and at the same time feel good about making a real contribution to improving our environment.

Resources

Organisations

Alliance for Beverage Cartons and the Environment (ACE)
☎ 0870 442 6000. See www.ace-uk.co.uk for interactive map

Black Environment Network ☎ 01286 870715
www.ben-network.org.uk

Centre for Alternative Technology (CAT) ☎ 01654 705950
www.cat.org.uk

Community Composting Network (CCN) ☎ 0114 2580 483
www.communitycompost.org

Community Recycling Network (CRN) ☎ 0117 942 0142
www.crn.org.uk

Community Re>Paint ☎ 0113 200 3959
www.communityrepaint.org.uk

Department for Environment, Food and Rural Affairs
(DEFRA), Nobel House, 17 Smith Square, London, SW1P 3JR,
Helpline 08459 33 55 77: choose option 5. www.defra.gov.uk
helpline@defra.gsi.gov.uk. Among many publications produced
by the DEFRA is an annual summary of environmental statistics
(*The Environment in your Pocket*), available on the website.

The Environment Agency, National Customer Contact
Centre, PO Box 544, Rotherham S60 1BY.
enquiries@environment-agency.gov.uk www.environment-
agency.gov.uk. The Environment Agency has a useful website
and produces a free newspaper, Environment Action, which
covers many water and waste issues in a lively way.

Federation of City Farms ☎ 0117 923 1800
www.farmgarden.org.uk

Friends of the Earth ☎ 020 7490 1555 www.foe.co.uk

REDUCE, REUSE, RECYCLE

Furniture Re-use Network (FRN) ☎ 0117 954 3571
www.frn.org.uk

Garden Organic ☎ 0247 630 3517 www.gardenorganic.org.uk

Global Action Plan ☎ 020 7405 5633
www.globalactionplan.org.uk

Green Stationery Company www.greenstat.co.uk

Money to Schools ☎ 01372 723723
www.moneytoschools.com

Recycle More ☎ 08450 682 572 www.recycle-more.co.uk

The Recycling Consortium ☎ 0117 930 4355
www.recyclingconsortium.org.uk

The Recycling helpline ☎ 0800 435576 will put you through
to your local council's helpline

World Wide Fund For Nature (WWF) www.panda.org

Waste Watch ☎ 020 7549 0300 www.wastewatch.org.uk

Womens Environmental Movement (WEN) ☎ 020 7481
9004 www.wen.org.uk

WRAP Waste and Resources Action Programme helpline
☎ 0808 100 2040 www.wrap.org.uk

Other useful websites

www.junkk.com Great fun and well worth a look

www.recycleworks.co.uk Actually about composting

www.greencone.com Food waste compost solutions

www.smartsoil.co.uk Food compost solutions for schools etc

For your local council, go to www.yourcouncil.gov.uk (e.g.
www.devon.gov.uk) or try an internet search using key words
like "Reduce, Reuse, Recycle" or the material you want to find
out more about. Alternatively, phone your local council and ask
to speak to a recycling officer or waste minimisation officer at
county, district, city, borough, metropolitan or Unitary level.